Plane Truth

Carl A. Davies

Plane Truth
—
A Private Investigator's Story

Algora Publishing
New York

Algora Publishing, New York
© 2001 by Algora Publishing
All rights reserved. Published 2001.
Printed in the United States of America
ISBN: 1-892941-57-0
Editors@algora.com

Library of Congress Cataloging-in-Publication Data 00-011415

Davies, Carl A.
 Plane truth : a private investigator's story / Carl A. Davies.
   p. cm.
Includes bibliographical references.
 ISBN 1-892941-57-0 (alk. paper)
 1. Aircraft accidents—Investigation. 2. Aircraft accidents—Political aspects-
-United States. 3. International offenses.
4. Terrorism. 5. Conspiracies. I. Title.
 TL553.5 .D39 2001
 363.12'465—dc21
                        00-011415

New York
www.algora.com

This Book is Dedicated To All Those Who Died In Airline Crashes

TABLE OF CONTENTS

Prologue: 1996-2000   3

1. Setting the Stage   11

    Aviation Engineering in My Blood —11. Coming to America — 14. Living It Up in Libya (1968 - 1973) — 17. (Getting) Out of Africa — 24. Eternal Paris — 30. Back to North America — 32. Navigating the Skies — and the Corridors of Cash — 35. What's Politics Got to Do with It? — 40. Getting Gaddafi — 42. Getting the Americans – Flying over Lockerbie — 45. Toxic Waste, and Wasted Efforts — 50. Yugo Riddles — 53. On My Own, Again — 55. Raising Franck Curtin — 57. Now What? — 59.

2. "My Dinner with Franck"   63

    Evaporating Prospects — 63. On the Road Again — 65. Frank Talk about Boeing 747's — 68.

3. Libyagate   79
4. Lockerbie, Scotland: Pan Am 103 Investigation   91
5. Everett, Washington: Birthplace of the Boeing 747   99
6. Gathering Evidence: Whom to Ask? Whom to Trust?   107

    Timing Devices, or Doubtful Evidence — 112. Drugs — 126.

7. TWA 800 and the Four Crash Theories   143

    The Terrorism Theory — 144. The Missile Theory — 145. The Central Fuel Tank Explosion — 148. The Metal Fatigue Theory — 152.

8. **Why Do Boeing 747's Break Up in Flight?** 155
Stress on All Sides — 155. Skin Problems — 159. A Little Help from Our Russian Friends — 160. Fail-Safing Successes —162. And Failures — 164. The Pan Am Report — 167. What Happened to TWA 800? — 169.

9. **Late-Breaking News** 185
UTA's DC-10 — 188. Fly, Britannia — 190. A Bit of A Scrap — Jim and Jane Swire — 190. Bollier, Or Odd Timing — 192. These are Trying Times — 203. The Lockerbie Trial — 203.

SELECTIVE BIBLIOGRAPHY 223

Prologue: 1996-2000

Figuratively speaking, I first put pen to paper on October 29, 1997, to document a story about the crashes of three Boeing 747 planes — Pan Am 103, TWA 800 and Air India 182. After a few months of investigation, I considered the story to have achieved sufficient drama to warrant publication. I was wrong. The story simply exploded in both content and intrigue, to the point where I constantly wondered when I could ever close the curtain on Act I.

At the time, President William Jefferson Clinton was surviving an embarrassing "Impeachment" crisis and was bombing Yugoslavia back to the 16th century. History will eventually conclude that the whole USA/NATO attack was an illegal act — not unlike the main thread of the story you are about to read. This story, or perhaps exposé is a better term, has no direct link to the war in Yugoslavia, but it documents a war nonetheless. The war began as an international economic war, but then the stakes began to grow exponentially and people began getting killed.

This war actually started in the early 1980's, but we will pick up the story on December 21, 1988, when Pan Am 103, on a routine flight from London's Heathrow Airport to New York, came to a catastrophic end over the delightfully tranquil and picturesque Scottish town of Lockerbie.

PA 103 was a Boeing 747-100, the fifteenth model of its kind; it "exploded" at 33,000 feet, killing 259 people on board and 11 people on the ground. The wreckage was spread over 845 square miles. Since then, more than twelve years have passed; many books have been written about the situation, any number of documentaries have been aired over television worldwide, Internet sites have been developed* and even feature movies have been made. Few, if any, have given a complete and accurate account of the events, and many "exposés" have been completely misleading. Indeed, many of the books and films have been hidden from the American and European public.

There is, however, a true story and it runs counter to what many experts have concluded and counter to the vast and often unreliable conspirators' analyses. There are many more aspects to this event than can be contained in any one book; our understanding of numerous stories of international business, war and politics will require new interpretations based on the revelations

---

* Excellent websites, including massive and comprehensive databases, are located at:
    1. www.geocities.com/CapitolHill/5260/
    2. http://headlines.yahoo.com/Pan_Am_103_Lockerbie_Trial/,
    3. http://headlines.yahoo.com/Full_Coverage/US/TWA_Flight_800
    4. www.TWA.com

These sites appear to be relatively accurate and certainly expose many points of view, including most of the theories and counter-theories of this unfolding situation.

in the current work. It is the author's intention for this book, and the movie associated with it, to serve as catalysts, stimulating the search for the few remaining missing links.

This book is the story of my investigation, rather than a dry research paper; others will write more detailed accounts, and historians will be analyzing these events for years to come. In 1997, I had the good fortune to meet Peter Hounam, an investigative reporter for the (UK) *Sunday Times*, who has written a number of books about world events — particularly those involving the more corrupt or sinister side of politics and business. I met with Peter many times, and he helped me get off to a good start.

While I was writing this book in the pleasant surroundings of Lake Neuchatel in Switzerland, and the seaside resort of Norfolk, UK, I regretted having to visualize Mr. Abdel Basser Ali al-Megrahi and Mr. Lemen Khalifa Fhimah in a makeshift jail in Holland under terms that would not be acceptable to Western suspects. Under Scottish law, these two men should only have been detained for 90 days before the trial began. For many years I have been convinced that these two Libyans were completely innocent, which triggered my dismay at their current situation. The overt purpose of the Lockerbie trial that was to begin on May 3, 2000 was to prove their guilt or innocence. As the trial unfolded, the prosecution suffered many setbacks in trying to prove their point — and hints of other, deeper plots came to light.

So many questions come to mind as we consider the known facts. Do we remember that on September 1, 1969, Colonel Muammar Gadhaffi orchestrated the military coup in Libya, deposing the ruling monarch, King Idris? And that King Idris was part of the Bedouin, Sennusi tribe (a name that will crop up later)?

Were Libyans truly involved in the April 4, 1986 "La Belle" disco bombing in Germany, as alleged? Or was it another group, as announced by the German police?

Do we remember that the Aegis Class cruiser, the U.S.S. "Vincennes," shot down an Iranian Airbus on July 3, 1988 killing 290 passengers some six months before the PA 103 catastrophe? It has been alleged that Iran paid $10 million to Ahmed Jabril, a former Syrian army officer who later headed up the "Popular Front for the Liberation of Palestine", to seek revenge for the downing of the Iranian Airbus.

Was the United States government shipping heroin from the Bakaa Valley in Lebanon for sale in the USA, to pay for the Iran-Contra affair, as alleged, and how might that have tied into the PA 103 bombing?

Could the theory of a terrorist bomb being planted on Pan Am 103, while it was in Malta, be a trumped up story? The Iranian baggage handling facility at Heathrow was immediately next to the Pan Am loading operation...

Can we be sure it was a bomb, at all? Bombs have exploded on many Boeing 747's and other aircraft but the planes landed, nonetheless.

Several events finally prompted me to close Act I at this point. At the time of this writing, the two Libyans accused of bombing Pan Am 103 were in a specially prepared jail at Camp Zeist in the Netherlands, which had been an airbase for the

American military. This base had been temporarily transformed into what was effectively and legally Scottish soil, on which was constructed a highly secured Scottish court, established for the duration of the trial of the two accused Libyans. Secondly, the United Nations' sanctions against Libya had been suspended. And thirdly, coincidentally, I had just finished reading a book by Michael Isikoff called *Uncovering Clinton*.

Why did this book make a difference to me? Issy, as he is known to his friends, was the reporter responsible for *Newsweek*'s winning the "National Magazine Award" on April 28, 1999, for his work exposing the now famous and embarrassing Monica Lewinsky incident to the entire United States.

Isikoff's award was based on "remarkable evenhandedness, clarity and close to complete accuracy" in reporting. In his book, he made it very clear that he was aware of the potential damage offered by the exposé and he assumed responsibility for the credibility of his sources. Now, I am a professional petroleum reservoir engineer whose responsibilities are to analyze the fluid flow in oil and gas reservoirs. Accuracy and honesty are as important in this profession as in reporting on a major story. I have discovered that analyzing oil and gas reservoir fluid flow behavior has very similar characteristics to analyzing the flow of human behavior. I kept these thoughts in mind as I gathered, assembled and committed these facts to paper for your consideration.

You may think I'm digressing from the subject — bear with me! I read Michael Isikoff's dust cover, and these emboldening remarks stiffened my resolve. I became obsessed with the desire to bring this effort to its rightful conclusion. I could not express my feelings better:

> In the end, Clinton's serial indiscretions really did matter. They mattered not because they had continued well into his tenure in the White House, and it mattered not only because, I was now convinced, Clinton was a far more psychologically disturbed individual than the public ever imagined. It mattered because private misbehavior on Clinton's scale required routine, repetitive, and reflexive lies to conceal itself. The lies were told at first by Clinton and then spread and magnified by everybody around him. The lies were easily rationalized on the grounds that it was Clinton's private life that was at issue — a matter of no public consequences. But lying, engaged in often enough, can have a corrosive effect. If lies were needed to avoid political embarrassment, then lies — or at least extremely mangled versions of the truth — would be told. A culture of concealment had sprung up around Bill Clinton and, I came to believe, that summer (1998), that it had infected his entire presidency.

Most people know something about the tragedy of PA 103; and some of the inconsistencies in the public story have come to light. Why did CBS's *60 Minutes* take ten years to investigate and show the American public another side to the Pan Am 103 story, and was it complete in every detail? Why is the entire American media ignoring such an important story where so many Americans were killed? Many more questions remain, and there is plenty of room for suspicion.

Prologue

I do not subscribe to conspiracy theories, I rely on facts. And I have tried to remember: "Every man is entitled to be wrong in his opinion; no man has the right to be wrong in his facts" (Bernard Baruch). The facts behind these disasters, and the stories that enshroud them, are far more complex than we have been led to believe. I intend to present to the reader various strands that may appear, at first, unrelated; and I promise that their conjunction will surprise you and reward your patient reading.

CHAPTER 1

SETTING THE STAGE

Aviation Engineering in My Blood

Why would a country boy from Welwyn Garden City, Hertfordshire, England, born of fine parents whose only desire was to live a modest comfortable life in a God-fearing world, be faced with exposing the true story of a major war, an economic war, taking place in the world today?

I intend to lead you through a series of events in my life, events that caused me to question certain tenets — especially those involving the news media, political figures, and causes — to give you the texture and the fullness of the investigation as it unfolded. (If you are not interested in the details that led me to take on this project, then I invite you to skip ahead and go straight to the heart of the exposé.)

My dad, Harry Davies, was brought up in an era when it was difficult to obtain an education; he ran away to Canada at the age of sixteen. Life in Canada in the 1920's was rather difficult, and he returned to England some seven years later. He became interested in the manufacture of airplanes, and has been a valuable advisor to me on aircraft metal fatigue and failure problems.

Harry Davies joined the De Havilland Aircraft Company, at Hatfield, Hertfordshire — at the time, a world leader in aircraft design and manufacturing. My Dad did not go to the front in the Second World War, the front came to him. He was building the renowned Spitfires and Mosquitos, which had not gone unnoticed by the German military. He used to cycle home for lunch, which saved his life when one day a bomb hit the factory (causing a great loss of life). He worked hard, long hours, and was dedicated to the firm. He eventually became an aircraft inspector and was responsible for making sure the planes were airworthy before the pilots were allowed to fly, and he signed documents to that effect.

These were early days in the business and many of the planes were experimental. Harry Davies signed off two planes, DH 110's, which had a very dangerous design and eventually killed the son of the founder, Geoffrey de Havilland, and John Derry. He had signed off the planes as meeting the manufacturer's specifications, and strapped in both pilots. As the last person to see these two pilots alive, he had the unfortunate task of identifying the bodies. These two deaths had a profound effect on my Dad's health. His long-term anguish was completely ignored.

The importance of this part of the story began in about 1949, when the de Havilland aircraft company designed and built the very first commercial airliner, the Comet. I intend to bring up the Comet problems later in the book, to illustrate further points. Harry Davies was the chief inspector on the very first Comet plane, and he was on the plane that made the fastest trip between London and Cairo in late 1949. He also crossed the equator at the highest elevation at that time; the trip was an experimental operation to Nairobi to check the effects of high temperatures on the plane's structure. The tail number (G-ALVG) of the first Comet airliner is clearly marked and a photograph is enclosed (page 15).

The Comet was an immediate success and airlines all over the world began placing orders. Then disaster struck. Two Comets disintegrated in flight, with the loss of all on board. The planes were grounded and, shortly thereafter, the Boeing 707 began service with great success. It took years of experimental work to locate the problem with the early Comet airliners. It turned out that the designer of the Comet, F.T. Hearle (Frank Whittle designed the first jet engine), after having enjoyed an outstanding career, decided that the windows should be made square, not to resemble to a ship's porthole. This was an unfortunate error. The corners of the windows caused a stress weakness, which ultimately caused failure in flight — a metal stress fatigue problem.

Once the problem was identified, a new version of the Comet was built. The Comet 3 was designed, built and flown on a world trip in an attempt to recover the market. My Dad was on the world trip, to Cairo, Bombay, Singapore, Darwin, Sydney, Melbourne, Perth, Auckland, Fiji, Honolulu, Vancouver, Toronto, Montreal and back to London.

After high school, or the "upper six" as we called it at the time, I was one of the few lucky students to obtain a place at a red brick university with a county grant. Fewer than 5% of the kids in those days (1958) had a shot at university in the United Kingdom. My happy days at Birmingham University, enrolled in the chemical engineering program, were enhanced beyond belief by the opportunity to fly in the University Air Squadron, which was a branch of the Royal Air Force. I enjoyed the chance to fly airplanes and even, on a limited basis, flying jet planes: the Vampire, the Jet Provost and the Hunter N7.

Thus, I learned to fly, and understood flight control and the mechanics of how these marvelous machines worked. My interest in airplanes was established. My Dad even arranged for me to ride on the Comet while John Cunningham, the de Havilland test pilot, took the plane to about 40,000 feet for stall tests. Hopefully you will never experience a stall in a commercial airplane, but if you ever do (and live to tell the tale), you will not forget the experience very quickly.

Coming to America

After graduation, I realized that working in the UK chemical industry was a waste of time. The pay offered by Exxon (known as Esso at the time, and the British all thought it to be a British company), 800 pounds sterling per year, meant that I would be subsidizing its development since it was insufficient to make a living. I decided that I was destined to be a fighter pilot. But that was not to be. I was suddenly offered a job by Mobil Oil

Setting the Stage

Above: Comet IV, note the round windows (Harry Davies, third down).

Below: Comet I, first airliner, note the square windows (Harry Davies, right under the nose).

to work as a petroleum engineer in Alberta, Canada.

I sold the 1938 Morris 8 for forty pounds, bought an engagement ring, proposed to my girlfriend, said goodbye to my poor parents and boarded the Boeing 707 for Canada. I was so eager to leave England that I did so with only sixteen pounds sterling in my pocket, which we consumed with joy over the Atlantic. I arrived in Calgary, accepted a $50 advance and was taken for a flight around the Alberta oil patch. I actually flew the twin engine plane and could not wait to go to work. That was the last plane I flew in Canada, even though Mobil Oil had indicated that I might be the equivalent of the Australian bush pilot. I was naïve and allowed myself to be conned.

Throughout my teenage years, I had wanted to go to "America"; to me, Canada amounted to the same thing as the US. Canada was a delightful experience. I had a great job as a petroleum engineer, working in the bush drilling oil and gas wells, installing pipelines and arranging for the products to be sent to market. I was enjoying the work and producing nature's bounty for all to enjoy. The country was wild, compared to my village in England, and I loved it. Janet and I were married in Edmonton and we soon had three wonderful boys.

There was only one problem with this great experience in Canada — the financial load was increasing daily. In 1968, we were transferred to Calgary, where I worked as a reservoir engineer writing computer programs on very early machines (IBM 1620's, for those who can remember these early wonders). We were very happy with life but literally living from paycheck to paycheck; or to phrase the concerns more realistically, one paycheck from disaster.

The oil business in Canada was very slow; but the international scene was hot, particularly in Libya. Many of our friends left Canada to work in Libya and the reports that came back to us were extremely attractive. I decided we needed a change. The children were very young and the idea of playing in one big sand box with excellent schools seemed quite acceptable to all. There was a veritable oil boom going on with major oil reserves having been found by Occidental Petroleum, a one-time bankrupt company from California run by Dr. Armand Hammer. Oxy, as it became known, had taken a lease from Mobil Oil and discovered a massive oil reserve (called the Inestre oil field) in the Libyan desert. Oxy laid a massive pipeline across the desert in record time and became a darling of Wall Street. His efforts in Libya changed the world, for his own good, and may conceivably have set certain events into motion that would ultimately lead to the suffering endured by the Libyan people some ten years later. The fact that oil prices rose significantly in 1973 was based on work by Oxy and, I believe, Arthur D. Little, the consulting firm headquartered in Boston, which had a contract to advise the Algerian Government on world oil and gas supply and demand.

Living It Up in Libya (1968 - 1973)

I interviewed for jobs with Mobil Oil, Oxy and the Oasis Oil Company, and was quickly offered a position with Oasis — very attractive, with a new office overlooking the Mediterranean Sea. Life in Libya was exceptional. We had more money than we had ever experienced and had little place to spend it (although a few trips to Beirut could quickly lighten the pocket, with wives buying up gold trinkets. Jan, unfortunately, did not have this opportunity having to care for three growing children). Our savings account grew each month and eventually bought us our first house in Paris. I learned to scuba dive at the American Air Force Base, Wheelus, and enjoyed the beaches, the weekend trips into the desert, the excellent weather and the parties. Expatriate life can be both fun and financially very rewarding.

Then came a major change — the end of the good life? On August 30, 1969, we came back from vacation with my cousin Jean and her husband Ray. Ray was a steward on British Airways, and all he wanted was a quiet vacation anywhere along the 2,000 miles of beautiful sandy beaches of Libya. He had very poor timing.

At about 7:30 the next morning, I was driving to my office at Oasis and had stopped for gas along the way. I noticed a lot of unusual army traffic racing along the Zarvia road, with soldiers all over the place. I asked the gas pump attendant what was going on and he recommended that I return to my villa. The September 1, 1969 Libyan coup, led by Colonel Gadhaffi, was underway.

I went back to my villa and spoke to Ray. He was more than unusually upset since he had been through a coup in Nigeria and had found it somewhat unnerving. We decided to investigate the situation. I picked up my Super Eight movie camera to record a piece of international history, and Ray and I took off in my VW to see what was going on. There was plenty of activity, with occasional gunfire (which was a little disconcerting).

I started my movie camera and begin recording the activities. Suddenly, out of nowhere, a soldier ran forward and stuck his automatic AK-47 in my ear. Fortunately, the soldier spoke English. (I understand the coup leaders had sent soldiers who could speak English into Georgeanpopoli, a suburb of Tripoli where most of the expatriates lived, to be in a position to calm any panicked people). Ray and I were told to get out of the car, hand over the camera and line up against a wall around a corner. We walked around the corner and were immediately surrounded by about twelve more amateur photographers, some of whom were in a complete state of panic and crying at the thought of their summary execution. I must admit that being lined up against the wall by soldiers was somewhat disturbing, but I knew that expatriates who behaved themselves and stayed calm were never badly treated in such circumstances — in those days. It would have been very poor politics to wipe out a bunch of expatriates. I would not be quite as confident today. Needless to say, our films were all exposed to the light; we were ordered to go home and stay there until further notice. It was curfew time.

Poor Ray, locked in our villa. Actually, he was a savior, having dealt with irate passengers on the airlines. He was extremely amusing and quickly helped us all get organized for what might be an extended period of hardship (especially with three young boys to contend with). The women prepared the first meal and were careful to keep the potato and carrot peelings, just in case we could not buy any more food. Ray soon found my long ladder and prepared a viewing stand on the roof of the flat-topped villa, where we had a great vantage point from which to observe the coup's progress. He served us drinks in his usual way, and we were very relaxed after a few gin and tonics, despite the tracer bullets passing overhead. We were probably a little cavalier about the situation — a few people experienced stray bullets coming through the windows. Most of the guns were being fired into the air, to create the impression of a major offensive. In fact, there was little to no resistance and only a few people were killed at the airport trying to defend it. Oddly enough, the Libyans who worked for Oasis had told us it could be expected months earlier, but nobody had taken any notice.

The next few days were an experience of a lifetime. How did Libya treat its resident foreigners — with whose Governments, for all intents and purposes, it was having serious political differences? The price of oil was beginning to rise far faster than anyone had experienced in twenty years or more, and that was viewed by the West as a hostile sign. What was going on?

The oil producing countries were hiring outside consultants to determine the balance between the supply and demand of oil on a worldwide basis. Such companies as Arthur D. Little, and Scientific Software, in Denver, were hard at work with very sophisticated computer models to make these determinations. I know, because I interviewed for a job with ADL in Algeria in 1972 and worked with SSC in Denver for many years while with the Oasis Oil Company. (This is another story for another day, but it is all interlinked in the global economic war.) Oddly enough, in Daniel Yergin's book *The Prize* (he is considered to be an authority on the recent history of global oil production), he never mentioned who was advising the producing countries. We all believe in free market forces and the producing countries had every right to raise prices if justified by the market demand. Furthermore, the US companies were conducting business legally and, looking back, they were providing the world with a very valuable lesson in business competition. The system has worked well, and today oil is cheaper than ever, if one takes into consideration the time devaluation of money.

Still, on this background, after the September 1, 1969 coup one might have expected to be treated by the Libyan forces as a hostile intruder. Nothing could have been further from the truth! We were guests in the country and we were treated as such. There were a few radical elements, quickly controlled, and no one was harassed or bothered in any way.

Life had changed, of course. Gadhaffi decide he was going to take his people back to their roots and all Western decadence was to be removed. In some instances, the move was amusing, sometimes wasteful, sometimes tragic — and many times it made no sense whatsoever. For example, the soldiers were ordered to remove any signs not in Arabic. Next thing we knew, the soldiers were shooting up all the signs around town, which made a huge mess. These damaged signs were left full of holes for the five more years that I was there. And all the English signs on the oilfield equipment were painted out; but the equipment was being run by expatriates who spoke only English. This made pumping the oil very hazardous, since the operators had no idea on how to control the equipment. This problem was later corrected using a painted code (not unlike a bar code), but at great expense, and it created a dangerous operating situation.

As for day-to-day living, we were treated with unbelievable courtesy. For the first week or so, we were under curfew all day and night, and we had no way to get food or anything else. We were told that anyone breaking the curfew ran the risk of being shot, so the choice was very simple. However, a wealthy neighbor across the street had a manservant, called a "Gaffia", who was about seventy years old and totally deaf. He would be sent down the street to the bakery with a large wicker basket on his back. He was very old for his age and hobbled slowly down the middle of the road with his aged body stooped over. Next thing we knew, this old man was coming up the street handing bread to all of the neighbors. After a week or so, the curfew was lifted for two or three hours each day for people to get groceries and other essentials.

At first, it appeared that we had a major problem. The banks were all closed, so no one had money to buy anything. If that situation were to occur in the Western world, one would need to plant a garden rather quickly, or learn to fish! Not so with the Libyan merchants. We simply gathered whatever we needed, then the merchant would add it up on a paper slip and hand it to us — making no record whatsoever. They trusted us. I have to believe that everybody went to the banks as soon as they opened and settled with these merchants. I doubt they lost a dime from the expatriates.

Within a couple of weeks, the curfew was lifted until dark, and we were able to take our houseguests to see the magnificent Roman ruins in Sabratha and Leptis Magna, which were modern, vibrant cities some two thousand years ago. The ruins in Rome pale in comparison to those in Libya. Leptis Magna is a complete town, and was the breadbasket of Europe centuries ago. Most of the land is a desert, but do not let that word fool you; it does not mean that there is nothing there. This is a fascinating country. Life in Libya during the next four years was very interesting. We could observe a country in its early development, while I had a good job and the children were being well educated in the local American school. Vacations and weekend travel to neighboring countries were added benefits to the expatriate life.

It must have been about September 1, 1973 when our family was returning from a holiday in Spain, only to be surprised by an announcement at the airport check-in desk that "something" was happening in Libya and the airline would not be responsible for us if we took the flight. I decided that if the planes were prepared to go to Tripoli, then we were prepared to go, too, back to where I could sort out the problems. That was a mistake.

We landed at the Tripoli International Airport and the whole family was taken into custody: for having entered the country illegally. This, I did not understand, since I had been working in Libya for nearly five years. It turned out that Libya had decreed that all foreign passports had to be in the Libyan language; clearly, our Canadian passports were not. The Canadian Government had informed Libya that its passports were in French and English, which was acceptable worldwide as set by a Geneva Convention. Well, Geneva Convention or not, Libya had its own rules and if you didn't like it you could go home. Or try to.

(Getting) Out of Africa

We spent all of that day and all night locked up in an un-air-conditioned room at the old airport with no drinks or food, with an armed guard watching over us (and getting nervous, as the boys kept chasing around). Then came a real harassment — or simply a stupid dilemma. I was told we were going to be shipped out the next morning to Malta and that I needed to buy airline tickets. I had no money — and could have been arrested immediately if I had had enough money for the tickets, since it was a crime to take out, or bring in, more than a certain amount of the currency. Credit cards were unacceptable. I was not allowed to use the phone. Crazy. I told the person in charge that it was his problem and I would just wait for his reply. A few hours later, I was allowed to make one phone call, in the presence of a Libyan who spoke English. Soon one of our friends, Howard Schlieman, came in with the airline tickets, and we were shipped off to Malta.

Malta is a delightful island north of Libya, and if you must embark upon the life of a refugee, I can highly recommend the Malta Hilton. We were certainly not too unhappy to spend the next three weeks on an extended all-expenses-paid vacation. On the other hand, I was getting tired of the nonsense, the children were missing school, and I was not sure if I had a job or not. Furthermore, all of my worldly possessions were now located in hostile territory.

We decided that enough was enough, and we were going back to Canada. Jan and the boys went to her folks in England,

and I eventually got back, after all the fuss, as though nothing had really happened. It must have taken me three weeks to settle my affairs, and since I was no longer working for the Oasis Oil Company (which had treated us all exceptionally well), I decided to do a little furniture and equipment trading to boost the final take-home pay.

I was never a really good student, and French was a particular strain on my intellectual capacities. I never understood why we needed to learn another language, since all of my friends were quite conversant in English. It took me much private coaching to get a passing grade in French. Then I was offered a trip to France on a student exchange program. I was amazed at how many people spoke French with the greatest of ease; a whole nation of them. I loved Paris, and that changed my life for the better, once more. And now, my French was making me money in the final days in Libya.

The Libyan leaders had first kicked out the Palestinians, who were quickly followed by Italians, then the Americans. They were being replaced by dozens of eager Frenchmen who, rightly or wrongly, believed that the Maghreb countries (Morocco, Algeria and Tunisia) should be extended to include at least Libya. Three of the four countries already spoke French as a second language. These French had no villas and no furniture, but they had ample cash. All it took to induce the Americans to return to the USA was to enable them to sell their sand-ridden washers and dryers for the same price they had paid for them over five years earlier.

For 10% of the take (and all of their bottles of booze), I was a willing interpreter. Word quickly got around and I became quite an industry, upsetting the oil company managers who were losing valuable staff to this French-speaking entrepreneur.

Then came an even greater coup for the Davies' fortunes. I had heard through the grapevines (e-mail in those days was d-mail: you "drove" to see the folks you wanted to talk to) that a brand new MGB sports car was sitting in the basement garage at Oxy's offices and nobody could make it work. The owner had died and the widow was prepared to take anything for the car, located in a place she had no intention of visiting, even if she had known how to fix it. Well, this country boy had rebuilt his first car engine at the age of twelve. A thousand dollars was the agreed price and the car was towed to my villa. I suppose it took me all of an hour to realize that one of the intake valve push rods was bent and there were any number of old MGB in the local junkyards with just the right part. It actually took me longer to wax and polish my new car than to fix the engine. Then, when it was sparkling new, I drove my first new car to Tunis — where the people speak French as a second language — just for fun.

On my return to Tripoli, I suddenly I realized that my new car would have to be exported! I fully expected that such arrangements would not be easy and indeed, it would have been impossible had I not acquired a large stock of booze (which was now banned in Libya) from the parting Americans. Well, everyone has an Italian friend and they like deals as much as anyone does, especially if they involve screwing some government, including their own. It turned out that Alitalia, the Italian airline, was flying to Kenya and bringing back fresh meat to Tripoli.

Normally the plane was empty on its return trip from Tripoli to Rome; it was not, however, the day I wanted to ship out the MGB. A little cash and a lot of booze was all it took to see my car fly to Rome. (Of course, I could not be sure that I would ever see it again.)

Meanwhile, I needed another job! The market for petroleum engineers was in Indonesia, Nigeria and other places that were not particularly attractive to a guy with a family. I was reading *The Herald Tribune*, while selling American refrigerators to the French (who had never seen such huge appliances before), and I noticed an advertisement for a petroleum engineer in Europe. Ideal!

I quickly contacted the Arthur D. Little office in London and was soon on my way to the interview. The job turned out to be in Algeria, to be preceded by a three-month stint in Portugal; I was extremely disappointed. I could well imagine Europe extending over the English Channel but not across the Gibraltar Straits, even if the distance is less. I was despondent at the prospect of entering another Arab country, and one that was experiencing similar upheavals to Libya's. My passionate desire to see the world, or incurable curiosity, drove me to accept a paid trip to Algiers even though I had no intention of taking the job. I needed a break anyway. Reading *Dollars for Terror: The US and Islam* (by Richard Labévière), I see that I was lucky to have avoided the unfortunate religious (or political) upheavals in Algeria in the mid-1990's.

ADL put me up in the best hotel in town, the Hotel Algiers; it was quite opulent but the beds were damp from poor air-conditioning. The next morning I was called to breakfast, and who turned up but Bob Williams — from whom I had leased my villa in Tripoli when he left Libya! We got along fine; the people were great and the job was fascinating.

It was a group of mainly British and Lebanese, doing computer modeling of the global oil and gas supply and demand. The work was being conducted under a $1.5 million contract with the Algerian government. An American company, working on contract, soon to be advising Algeria and Libya to hike the price of oil and gas! I was told later that they were also working with a fine gentleman called Taki Rifia, whom I would meet at the Paris-based bank, "Banque Arabe Internationale." It was the spring of 1973.

I flew back to Libya to finalize plans for my departure. I went back to selling furniture to the French. Then the strangest of events was to occur. I was leaving my office for the last time, and walking out the door, when the phone rang. I was tired of being a furniture salesman and wanted to relax for a few days but curiosity, again — or the prospect of a little more cash — made me go back and pick up the phone. It was Bob Williams, from Algeria. He was very excited. He told me that he had quit ADL and was off to Paris, to start an Energy Department at the Banque de la Société Financière Européenne. He had noticed that I spoke French, in Algiers, and thought that I would be an ideal colleague to set up the department. I was overjoyed. I left for Paris, made my deal with the bank, then went back to Libya to complete the departure process.

We were not permitted to own property in Libya, so all the

expatriates were obliged to rent their villas. The company always drew up the leasing contracts, in Arabic on one side of the page and English on the other. My landlord's only obligation was to maintain the building in good order and mine was not to destroy it. The very first time we had a heavy rain, the roof had leaked. I promptly informed the landlord; he ignored that request, along with many other repairs, and continued to do so for the next five years of the lease. He knew we would fix things ourselves, rather than have the rain dripping on our furniture. I fixed the roof by laying sugar sacks over the holes and pouring tar over the sacks. It worked very well, even better than the so-called professional jobs. I also drove a nice round hole in the roof for the potbellied fireplace I had bought from Sears Roebuck. Among other things, I also cut a rather large hole in the double doors to install a swamp cooler. (If you are unsure what a swamp cooler is, call someone from West Texas — they will know.) So, when I left, the roof looked like a patchwork quilt with a 12-inch hole in one corner, and the door had been closed with cheap plywood. Needless to say, the landlord quickly approached me, with two "enforcers". The conversation about who owed what to whom was a little disconcerting. Then I remembered the way Jan would bargain with the gold and rug merchants in the Tripoli *souk*. I emptied my pockets of all the cash available and told him that was all I had in the world. He seem satisfied that he had cleaned out my worldly possessions (but not altogether happy). We had settled.

A friend took me to the airport and I boarded the plane. I finally felt relaxed; all was well, I was off to Paris to be a Vice

President of an international bank and catch up with the family. It was a great feeling as the plane began accelerating down the runway, ready to leap into the clear blue Libyan skies. Suddenly, the engines went into reverse and the brakes were applied with riveting force. I was terrified. I thought my landlord, who was an influential man, had stopped the plane and that I was going to be taken to jail. It turned out to be only a faulty instrument, so we all simply sweated in the plane for several hours while it was repaired. After we took off, the next time, there was a joyful cheer when the pilot announced that we were in international airspace: everyone was so pleased to leave the political tension for a well-earned rest.

The Libyan coastline disappeared over the horizon and I felt a curiously sad emotion of leaving this home for good. Libya had been a wonderful experience, we'd had many good times there, and met great people with whom we have maintained contact. I felt sorry for the Libyan people, most of whom were just beginning to extricate themselves from living in a third world environment. And I, lucky once again, was off to Rome to see if the Italians had been joy riding in my new MGB.

Eternal Paris

I must admit I was quite surprised to see the car sitting in the Customs yard, even though it was filthy, had no gas, and all the tires were flat. (The pilot had let down the tires so they did not blow up in the rarified air across the Mediterranean.) I hired a cab to round up a tire pump and a few liters of gas. Then, needless to say, I was soon racing with the wild Italian drivers (I love the Italians, and their country too) to Paris, France.

I could not believe my fortune! Here I was in the middle of Paris, working at one of the fanciest business addresses, 20, Rue de la Paix, near Place Vendôme and the Ritz Hotel. I traveled the world, first class, meeting top executives from the most progressive oil companies in the world. Our children were at the American school of Paris and we lived in a new American-style home in an ancient village called St. Nom La Breteche, about fifteen kilometers west of Paris. I quickly adapted to the life of a banker, even though entertaining my American oil company executives with fine cuisine was having an adverse effect on my finely tuned body. The weekends were a constant adventure into the French culture and the Banque Nationale de Paris's chateau with its beautiful grounds and private tennis club, in Louvecienne. This unique experience made me aware of the fate of the privileged classes during the French Revolution — the guillotine, for such opulent bankers! The pleasure was such that I rapidly recovered from any feeling of guilt. Working in Paris meant we were on everyone's list of eligible people to visit, and we welcomed all of our many visitors to the wonder of Paris life.

After about four years, a law was passed whereby each company was obligated to form a "Comité d'Entreprise", the equivalent of a union for each individual company. I was chosen to represent the foreigners in the bank (who were probably in the majority, this being a consortium owned by eight major banks). I was soon singled out as having my children in a private school, whereas everyone else's children went to state schools. This school was costing the bank as much as my salary — that was the deal I had made, and the bank was obliged to honor it until I left. The good old days were coming to an end. My children were not sufficiently well-versed in the language to go to a French school; and the bank was facing a weaker financial outlook, which put even more pressure on me to reduce costs. Luck was on its way.

Back to North America

One of my most promising clients, Grady Harrison, was a sophisticated entrepreneur from East Texas; he'd recently made a lot of money financing one of the first oil production platforms in the North Sea, with German tax-shelter money. Now he was raising large sums from the same sources, for drilling in the USA. He was also looking for bank financing to develop any oilfields he might discover; and that is why he looked to our Energy Department.

One day, Grady came to the office all excited. He had made a major oil discovery in Montana. His geologist had discovered wells with hundreds of feet of oil sands. I told Grady I was amazed to hear the story, since I was reasonably familiar with the Montana geology. We called a friend of mine who had recently left Libya to work in Montana, and he confirmed my suspicions. Grady's face went quite gray. He had announced the big find to the Germans, who were now raising more capital at an ever-increasing rate. The crooks with whom he'd been working had hit a high-pressure water zone, and they were shooting the water into the air, at dusk, when they took the photograph. Only upon very careful examination could we see that the jet of fluid was water, not oil. In the poor light, the jet looked black, at first glance. I told Grady that the days of allowing oil wells to shoot into the air had ended in the 1930's.

Grady came to the office about a week later to explain that he had cut off the funds to these crooks and filed a lawsuit against them. He went on to say that while he was one of the best at raising capital, he knew nothing about the oil business. He made me an offer I could not refuse. Jan cried at the thought of leaving Paris.

A month later, I was working in Houston on a contract basis. Grady hired an excellent German attorney, Angelica Lange (who had a license to practice law in Texas) to seek a "Green Card" to allow us to live and work in the USA. It is probably easier to wade across the Rio Grande River. Suffice that the coveted Green Card was issued.

I am going to recount a few of my experiences as a promoter, a deal-maker, an entrepreneur, a merchant banker — whatever you wish to call those individuals who work independently, using only their wits. The opportunity to see hundreds of deals all over the world was an important part of my development and helped me put together the pieces of this story.

Grady was a highly effective entrepreneur, and the modest success of our first drilling program could have been easily leveraged into a better deal had we merged with a public company.

Don Leiderman was a successful Jewish deal man who operated from a palatial house and office in Beverly Hills, California.

He had gained control of a company called Capital Energy, and, although he knew nothing about oil and gas, he was in the business. One day at lunch he said, in jest, that he never understood what the big fuss was all about when it came to finding oil and gas. When he wanted oil and gas, he had any number of gas stations almost giving the stuff away within two miles!

Don was good, but Grady was better. We soon merged our efforts into his company and we took stock in his. We went home to do something else, and left Don with all the headaches. It was a great deal for me; I was fortunate to bail out when the stock was at its highest. When Grady and his partner began to sell, the stock fell precipitously. I learned a good lesson: it does not always pay to be the big guy. Grady then began to be very irritable and demanding, and eventually I left the company. I should have stayed at least another year. Grady had been very good to me; he fetched up with leukemia and was dead within a few months, at less than fifty years old. Had I not have met Grady, this work would never have been undertaken.

One deal that had made Grady very mad was with the notorious Alan Bond. Bond got his start as London taxi driver, then emigrated to western Australia to become a house painter. In Perth, he developed an ingenious long-term plan to make himself extremely rich. He became a successful real estate developer, but he was also an avid sailor and loved to race big yachts on the Indian Ocean. Mr. Bond had his eyes on the America's Cup, which had been challenged many times since the first race on August 8, 1870, but had never left American soil. Bond's idea was to win the America's Cup and bring it to Perth. Then, when all of the challengers and their admirers came to Perth, trying to win it back, he would sell them a place in his new village. Brilliant!

I was quietly working at Grady's office one day when our attractive secretary told us she had met Alan Bond, while moonlighting (selling jewelry) in the Houston Galleria shopping mall. She said Bond had bought thousands of dollars worth of watches, and wanted to take her out. Next thing we knew, Alan Bond was in our office making a deal.

He was in Houston trying to raise $35 million to take over an Australia gas company, Santos. Grady was not too keen, and he declined to participate. I liked the deal and agreed to go to Australia to put the deal together, if he paid my expenses.

Navigating the Skies — and the Corridors of Cash

Bond agreed, and I left for Perth, Australia. I had a great time flying all over the continent with Bond and his entourage of 10 or 15 hangers-on. It took two weeks to put the deal together with the help of two engineers I had known in Libya (Ray Hollis and Mike Babinak), who were now working for Santos.

Bond and I took off on a Boeing 747 to Zurich with the intention of meeting a banker friend of mine. Alan and I were in the upstairs cabin and the captain was very pleasant, knowing he was carrying the one and only Alan Bond (known as "Bondy", to his friends). We flew over the Indian Ocean late at night, and I asked the captain if I could spend a little time in the cockpit — being a would-be pilot, myself. I was particularly interested in the inertial navigation system since we were crossing a mighty ocean and needed to arrive at Bombay to refuel. These were the days before satellite navigation systems, now known as GPS, Global Positioning Satellites. The captain made it quite clear to me that they knew where they were, to within a few thousand feet. This is important to remember when we discuss another Boeing 747, the Korean 007, which was shot down by the Russians on August 31, 1983.

Alan and I landed in Zurich and met with Mr. Saul Marias, a friend and an impressive merchant banker. The two gentlemen did not see eye to eye, and Alan was furious that I had hauled him halfway around the world for such a reception. (Saul had checked out Alan's reputation and was not comfortable working with him.)

I told Alan we had alternative arrangements, and we were soon on our way see my friends at the International Energy Bank in London. Gordon Ahalt, the President, an excellent banker who was also a petroleum engineer, welcomed us warmly. The Paris bank for whom I had worked, BSFE, owned 20% of IEB. It was clear a deal was going to be made. I called Grady, whose attitude was now quite positive, at his Munich office. I was now a golden boy.

After the meeting, Alan took me back to the hotel in his chauffeur-driven Rolls Royce — and told me his was not going to pay the commission he had agreed to in writing. He said my fee was too much, considering that he had paid all of my expenses and suffered such indignation in the presence of Mr. Marias. I called Grady in Munich for advice, at about 10:00 that evening; he went ballistic. First he fired me and told me to go home the next day. Then he called me every fifteen minutes, until about 3:00AM. I had just arrived from Australia and was not sure if I were in Perth, Houston or somewhere in between. One thing was certain, my mind was not in London. The final call came at about 4:00AM and I was instructed to call Alan Bond right then, and make him sign the agreement again with International Energy Bank as our client.

I was terrified. I was terrified of Grady and I was quite sure that Bondy was going to go nuts, too. He did. Imagine being as rich as Bondy and having a stranger wake you up in the middle of the night, making demands, after what we had just been through.

This was his reply. "Carl. You have dragged me halfway across the world, had your friends insult me in Zurich, and now you are insisting on a meeting to change our agreement in the middle of the night. This is what is going to happen: You will be waiting outside my door at exactly eight o'clock in the morning. If you are there one minute early and I see you, the whole deal is off; and the same thing will happen if you are one minute late!"

I called Grady to tell him of my orders. That was a mistake; he took it as a personal insult. I was fired again, then told to be at the meeting as ordered, and to call back immediately once the agreement was signed. I was verbally abused for another thirty minutes before he calmed down. One tries to be philosophical... I needed the job.

I hid behind a corridor wall until my watch read 30 seconds to 8:00AM, and sure enough Bondy opened the door at 8:00 sharp and let me in. He was very pleasant, at first, then he grabbed a paper napkin and wrote "I, Alan Bond, will pay Carl Davies $450,000 US dollars if I close the deal with the International Energy Bank". He signed it and told me that I need not be further involved.

Grady threw a fit! He did not like the agreement being in my name, which I can understand, but Bondy would not sign it in the name of someone he did not know. Grady screamed at me for an hour, all the way from Munich. He was so loud I am confident I could have heard him without paying for the long distance telephone call. I was fired again (and pleased about it, this time). I was sure he would be happy to have $450,000, when the time came, and I had only worked two weeks. It was not right — but it was better than a poke in the eye. As it turned out, he was not pleased, and had another tirade when we met back in Houston. He wrote an ugly letter to Bond, stating that he would see him in court. Then leukemia took over the narrative; Grady did not see anyone in court, and needless to say, we never got a penny. A few years later, I found out that IEB had lent his company $235 million; our commission would have been over $5 million!

Bondy was on a roll. I swear my deal gave him the cash to build the Australia II yacht, with its controversial winged keel. In 1983, he took the America's Cup to Perth, with great fanfare. His real estate deal was the resounding success he had foreseen, and he was a national hero in Australia — but the politicians hated him. Bond then got too big for his boots. He eventually raised over $5.5 billion, then the balloon burst. Bondy landed in jail over some trading scandal; he was released for awhile then thrown back in over another scandal; today, I believe he has fully paid for his sins, and I am sure he has plenty of cash stashed someplace.

Needless to say, I was on my own again. I had met another effective promoter named Robert E. Thorpe, who had indicated that we might work together raising capital for the oil business. I was now reasonably well-versed in the business and I quickly agreed to a salary of $5,000 per month and all expenses paid. I opened an office in the Galleria area in Houston and began flying all over the USA in search of deals and funding. I received the first two salary payments, and then suddenly Mr. Phillip S. Fogg, Robert's partner, refused to pay my expenses, which exceeded my salary payments by nearly $8,000. I was dumbstruck! How could I have been so idiotic?

I sat in my new office with only a phone and one yellow pad. I had enjoyed the entrepreneurial life and was not eager to return to work for a major company.

I called a guy I knew at the bank in Paris, Taft Simons. He had never liked me. I asked him if he could find me work; he was not interested in me at all, but he did suggest I call Mr. E. Barger Miller. Barger was very surprised by my introduction but, being an anglophile, decided he would give me a chance. He handed me an oil project to fund. I had no idea how to go about the business, but I was very pleasantly surprised to find that oil people were always interested to look at well-prepared oil prospectuses. Within two months, we had over $5 million committed to four wells in the East Texas Cotton Valley oil and gas play. I was off and running.

The late 1970's and early 1980's were great times for oil

deals. We made money. Even my tennis coach, who would not recognize an oil wellhead if I sat him on it, was selling oil deals.

I suppose it was Christmas, 1982, when I first realized that the bubble had burst. I could not sell the last twenty five percent of what I thought was an excellent deal. I took the deal myself, only to find out that the driller had absconded with the money. Another hard lesson! There was worse to come. The word around the oil patch was, "Stay alive 'til '85 (1985), back in the chips in '86, we'll be back in heaven by '87". I believed the saying; by 1997, I could not understand why I had been so convinced. Ten years later, the oil business was still trying to engineer a turnaround, based on better techniques.

What's Politics Got to Do with It?

The 1980's were the Reagan/Bush years; 12 consecutive years of a growing economy, a very pleasant change from the disastrous effort by Jimmy Carter with his 21% interest rates. Then in 1986, while the idiot Democrats were still controlling the House and Senate, Dan Rostenkowski, the Chairman of the all powerful Ways and Means Committee, enacted the worst tax package ever forced on a gullible, American public: the 1986 Tax Reform Act. Rosty could not stand by while a Republican president, Reagan, enjoyed such a good economy. Rosty was quite prepared to force the economy into a tailspin, crushing the middle class so that he could impose his socialism on the unwashed, downtrodden, hapless American population.

The 1986 Tax Reform Act has cost the American taxpayers

at least $500,000,000,000 and probably will cost $2 trillion over the next twenty years. This initiative caused the real estate market to crash, which caused the Savings and Loans companies to go under, which caused the Texas banks to collapse and require bailouts by the Eastern banks, which caused havoc with the Texas economy. The American economy is so huge and amazingly resilient it can withstand almost any abuse from any government. Rostenkowski was finally caught with his hand in the cookie jar, which landed him in jail. Now, as I write, he has been released on probation with an annual salary of $130,000 of American taxpayers' money, for life.

I should have moved to New York, where the stockbrokers were getting rich, but the kids were in Texas colleges and it just did not seem reasonable to move. It seemed at this point that the world was against me. Nobody would hire a middle-aged petroleum engineer, so I decided to join the enviro-wackos. The public had been convinced that the world would come to an end if we did not clean up our environmental mess. I have always been an environmentalist (without knowing it), and now the money was flowing freely in this massive new market. In actual fact, at least 60% of the money has gone to the legal profession — which has not cleaned up one gram of the pollutants. I banked on my education in chemical and petroleum engineering, and plunged into the environmental business from both ends: the clean up of toxic and hazardous wastes and the upgrading of crude. We acquired some unique enzymes from the University of Houston to use as a natural method of upgrading heavy, sulfurous oil and remove pollutants from contaminated oils.

Getting Gadhafi

One nice thing about writing books is that one has the liberty of backing up in time, and that is precisely what I intend to do.

As you will recall, I spent five years in Tripoli, Libya between 1968 and 1973. Libya is one of the most interesting countries in the world with its ancient history, pleasant people, and beautiful beaches along 2,000 miles of unpolluted coastline. It could be a paradise. The country has enormous wealth in oil and gas, minerals, water, agriculture, sunshine, and tourism. I have been very disappointed by the waste that has taken place by the Revolutionary Command Council headed up by Colonel Gadhaffi and Jaloud. The world does not know the rationale behind Libya's idealism in fostering Pan-Arabism, since our access to information is filtered through the media — which has been less than honest in its reporting. Politics always comes into play.

It may have been the access to huge amounts of oil money that allowed Gadhaffi to build his army and fund crusades in Sudan and Chad, to subsidize the IRA and others. We mustn't ignore the fact that the arms merchants were all too eager to supply Gadhaffi with whatever he needed, as long as he had the cash. The case of Mr. Edwin Wilson is worth considering: he shipped plastic explosives to Libya and was sent to jail for it. Wilson claims he was ordered to do so by the CIA, but they, naturally, deny it. New evidence is coming to light about the involvement of the CIA, including a *Houston Chronicle* article dated January 20, 2000.

Gadhaffi created any number of unlikely Arab unions with Syria, Egypt, and Sudan, none of which amount to a hill of beans. These unions cost Libya huge amounts of money and a loss of prestige. I believe Gadhaffi, in his early years as a leader, was consumed by his interpretation of the injustices imposed on people in the under-developed world. In some ways he was right.

As everybody knows, the problems between the Israelis and the Palestinians began in 1948, shortly after World War II, and were caused when the British, French and Americans decided that a Jewish state could be created on what appeared to be rather poor land owned by the Palestinians. The Israelis, with the infusion of massive amounts of American foreign aid, brought about the development of a new, thriving society (albeit a socialist country, many of whose values are contrary to the beliefs of most Americans) on top of the Palestinian society.

I will spare you the details of my understanding of Israel's problems with the Palestinians — the problems have been going on for 50 years and the Western press are always bulging with Israeli hard luck stories. But Gadhaffi was passionate about the Palestinian plight and has spent untold millions on the cause. This has not gone unnoticed by the Jews, who wield a disproportionate amount of power over the Washington elite.

I have never read any proof that the Libyans were responsible for the bomb that went off in a German nightclub that was frequented by American servicemen. At about 1:30AM on April 4, 1986 a bomb did go off in the crowded *La Belle* discotheque in West Berlin. That was a very crude and cruel way to exploit one's grievances; several Americans were killed.

Libya was immediately held responsible, since it was considered a terrorist nation (along with Syria, Iran, Iraq and others). In retaliation, just ten days later, Ronald Reagan and Prime Minister Margaret Thatcher authorized a bombing raid on Tripoli. At precisely 17:36 on April 14, 1986, twenty-four F-111's were launched from the Lakenheath US Airforce base in Norfolk, UK. It would be amazing if the US could have fully analyzed the situation in such a short period of time, and justified and organized an assault on a sovereign nation. Clearly, this was a deliberate but abortive attempt to assassinate Gadhaffi, or at least to send him a very loud message.

The trip was extremely dangerous for the pilots. France and Spain were opposed to the raid, and they refused to allow the bombers to cross either country's territory; this meant that the pilots had to fly around them and enter the Mediterranean basin south of Gibraltar and north of Morocco, through the Straits of Gibraltar. The dangers were manifold. The planes required refueling four times over the course of the 18-hour flight and to make matters even more hazardous, the operation took place at night. In addition, there was the possibility of early detection as the planes flew down the Straits of Gibraltar and into the Gulf of Sirte. The Libyans were not short of radar equipment and had the ability to effect an interception.

The raid was deemed to have been a success, both technically and politically, since Gadhaffi has not mounted any known terrorist act since. Unfortunately, innocent people were killed, including Gadhaffi's young adopted daughter. For those readers interested in the details, there is an excellent account in *Raid on*

Gadhafi, written by the pilot who directed the attack, Col. Robert E. Venkus. This was one more foreign adventure by a country most of whose citizens cannot locate Libya on a world map; most people do not see the relevance of these events and the story does not rate significant media coverage.

The truth about Libya's involvement in terrorist activities will probably not be known for decades. What is certain is Gadhaffi's futile crusades into Sudan and Chad. Whether Syria, Egypt, Iraq, and Iran were part of the terrorist acts allegedly carried out by Libyans is still unclear. Certainly, in their minds, these peoples and their countries were not carrying out government-sponsored terrorist acts but a war against what they deemed to be unjust treatment by Western nations with a bias towards Israel. The excursion into Chad was particularly painful for the Libyans; they lost so many young men. By the end of the war, a front-line base was overrun and the West had easy access to billions of dollars of military equipment that had been supplied by the former USSR.

Getting the Americans – Flying over Lockerbie

The next important event occurred early in the evening of December 21, 1988. Pan Am 103 had left London's Heathrow Airport on a routine flight, taking many American kids home for Christmas. What happened on that fateful flight can best be described with a reference to my own experience flying home from Paris on January 23, 1997, after having spent three months on this investigation in Europe.

I had finished the first of my investigative trips and was heading home. I got up at 6:00AM in the Montana Hotel, Geneva, and rushed to the airport to catch an Air France flight to Paris, where I expected to connect to the Houston bound flight later that morning. The incoming plane did not arrive at all. Eventually, we were told there was a flight to Chicago, from where we would meet a Continental flight to Houston.

This was going to be a long day. I used to love going to the Charles de Gaulle airport when it was new, some twenty years ago. Now, the French park the planes out on the apron and have you scramble down the ladder with all your computers, printers, and phones, fifteen pounds of transformers, and the files that you cannot leave to chance with the baggage handlers. Then you are herded onto a bus, which goes on a wild tour of the airport before dumping you on the ground in front of some dirty door. At this point, you are faced with miles of corridors, with an unkindly series of stationary stairs both up and down.

Fortunately, the French will sell you a very powerful coffee that is guaranteed to pump your heart hard enough to effect a rather rapid recovery from the disembarking process and prepare you for an equally arduous reboarding ordeal.

Suddenly, there was a moment of absolute terror. I got off the bus and was faced with what appeared to be over 200 steps up the ladder to board. . . a Boeing 747-200. Oh, no, I thought: I am going to uncover a serious problem with the early-vintage Boeing 747's and die in the process. I very seriously considered not getting on the plane. How ironic it would be if I were to go down over the Atlantic, with all my documentation going to the bottom with me.

I took several photographs of the plane, including the tail number, so I could look up its history if and when I got home. A technician confidently told me that Air Chance serviced their planes frequently, with no expense spared. That was little comfort, but I decided that for this plane to fall apart on this trip would be inconceivably bad luck.

The plane took off late, at 1:10PM, and began its climb towards Scotland. I followed our progress across the world map, on the flight-monitoring video. We were clearly following a set course, which all planes follow on their way to North America. It goes over the town of Lockerbie, Scotland. As we approached Lockerbie, the plane was flying at 33,000 feet, or 10,000 meters, at a speed of 552 mile per hour. The outside temperature was 58° Celsius (-72° Fahrenheit). As we passed over Lockerbie, the plane went through some serious turbulence and my water glass fell into my lap! My heart bounced in my chest. I could just imagine this plane shuddering and wobbling as it came apart at the seams.

It must have been about the same experience for the passengers of Pan Am 103, until all hell broke loose over Lockerbie. The official story is that a bomb had been placed on the plane, and was triggered by a barometric and or timing device. The crash caused one of the most thorough investigations of all times into an airplane disaster. The FBI were on the scene within just a couple of hours, legitimizing their presence by the fact that it was an American plane (although, legally, it had crashed on British soil and was therefore under British law and jurisdiction).

Parts of the plane were spread over 845 square miles; but two very important facts have received very little attention in the printed press. One is that the nose section of the plane landed first in a field near the wee town of Tundergarth, opposite a very old and unique church. The second important point is that the FBI (not the British authorities) found a small piece of metal, allegedly from a Panasonic radio that had contained the explosives. It was said to hold traces of a plastic explosives called SEMTEX, which was originally designed in Hungary. Some five years later, another important piece of evidence was found in an isolated wooded area; it was a piece of the timing device for the bomb, which allegedly triggered the explosion. Five years later. . .

There are so many reference works on this disaster, so many books, movies, and PBS documentaries, that the author hopes to establish a public reference center like the one that was opened in Dallas after the John F. Kennedy assassination.

In April, 1992, the United Nations, with the wholehearted support of the USA and Britain, imposed the most severe sanctions on Libya. No flights were allowed in or out of the country; any country caught dealing with Libya could not do business with any of the UN countries — and above all, not with the USA. The sanctions were initiated on the basis of an intense international investigation that seemed to indicate that two Libyans had instigated the bombing of Pan Am 103. Their names were Lemen Khalifa Fhimah and Abdul Basser Ali al-Megrahi.

They were immediately added to the FBI's list of the ten most wanted men. Their pictures can be seen on the Internet. Gadhaffi refused to accept the results of the investigation and refused to turn the men over to any court in the USA or England (or anywhere else, for that matter). The hunt for these alleged terrorists and the sanctions that were tied to the incident went on for over five years, causing unbelievable hardships to 5,000,000 innocent Libyans.

The entire world was apparently furious at this wanton act of violence; everyone who flew regularly became very nervous, hoping they would not be the next victims. In fact, Americans were warned against traveling to Europe, which had a certain negative impact on the travel industry. But the impact was far worse on the Libyans, who ran short of medical supplies, food, equipment and other basics that we all take for granted. In order to leave the country, one faced an arduous 100-mile road trip to Tunisia or Egypt, or a long boat ride to Malta.

In fact, the sanctions against Libya were coupled with a similar sanction act known as the Libyan-Iran Sanctions Act, passed by Congress in 1996. It was also know as the (Alfonse) D'Amato Bill. There is a very interesting article in the "The Washington Report on Middle East Affairs", by Colin McKinnon, dated March 1998. McKinnon's article states that the Bill was drafted by the AIPAC, or the American Israel Public Affairs Committee. Why was AIPAC writing a document concerning two unassociated sovereign nations, a document that would have such a massive impact on so may innocent people? The US Government has over three million bureaucrats; surely some of them could have done the job. Another story for another book.

Toxic Waste, and Wasted Efforts

Back to May, 1987, when I decided that the future of the petroleum engineer was already in a museum. According to the California enviro-wackos, it was "good riddance to a polluting race of parasites, go find a proper job!" They made this claim even as they burned millions of gallons of gas in the California freeway traffic jams.

I became a supporter of the environmental movement since that was where I thought I could make a living. I was introduced to a lawyer turned entrepreneur, Thomas A. Dardas. He had taken over a company called Detox Industries, Inc., which had a technology for biodegrading polychlorinated biphenols (PCB's). For those of you who invested in Detox, please keep reading my book (I need the money) — I, too, lost my shirt on Detox.

I am going to spare the reader the agony of that fiasco, but let it be said that while Dardas and I had many business fights, they were only borderline personal. For example, he named one of his computer files "Carlshit". Dardas is of Greek origin and has the temperament of a Mediterranean on a hot summer's day. Dardas is a greedy individual and greed is good — right? He is a lawyer and excellent with words; he would have been in his element during Clinton's impeachment trial. "That depends on what the definition of 'is' is." Dardas would convince you of certain facts only to cleverly twist them to his own advantage, later.

I was a struggling entrepreneur, trying to make a living in the business war zones. Forget alligator alley, business is war and it takes no prisoners. I was also an engineer, a banker, a fin-

Setting the Stage

ancier who (despite a lack of real success in business) was accumulating hundreds of contacts. One day, a very good friend called to say that a company out of Phoenix, Arizona was trying to raise about $350 million to build a commercial airport and a commercial plane refurbishing operation to meet a huge anticipated demand. I knew one private company in Birmingham, Alabama, that was in the business of building airports and was loaded with cash. I decided to take a look at the deal.

Sometime in that spring of 1989, I left Houston for Phoenix and was met by Larry Bell at the airport. Larry was a successful real estate broker and was working with a Frenchman named Franck Curtin, President of Trans-Atlantic Aerospace, Inc., to acquire the necessary land for the operation. I signed the 3-year non-disclosure agreement at Larry's office and proceeded to read the business plan. A few hours later, Monsieur Curtin came to the office to see who was this guy who thought he could raise the capital, when he himself was having so much difficulty.

Curtin was as cool as any Frenchman can be with a stranger. I had lived in Paris and loved the atmosphere, so the typical French indifference to foreigners had little effect on me. I remember saying a few words in French, as we motored out to the Buckeye Airport just west of Phoenix; my sallies were either ignored or responded to in English. Well, I had tried to break the ice, had I not?

Buckeye Airport was nothing more than an old World War II concrete slab in the middle of the Arizona desert. There were a few propeller planes doing "circuits and bumps" to borrow a British air force term for going around and around the airfield taking off and landing for practice. These were the next class of Lufthansa pilots, Lufthansa having found it cheap and effective to train its potential pilots in the clear desert skies of Arizona.

The deal looked excellent. Franck Curtin had spent over $1.5 million dollars on the business plan, which was amongst the most thorough ones I had ever read. This Trans-Atlantic Aerospace document is a major part of our story and it will sometimes be referred to as the "Boeing Report." After a brief discussion with Monsieur Curtin, I believed he had some confidence that I could perform. I was given a copy of the business plan to take back to my own office. I was intrigued.

This was a massive undertaking. The business plan listed all the major commercial aircraft in operation all over the globe. Every major airline was included, with a list of its currently owned planes, mainly Boeing airplanes of the older vintage (727's, 737's and 747's), which were detailed for extensive repairs. There were diagrams, prepared by Boeing, on all the repairs that would be required, together with a timetable for the work. These were all structural repairs, not normal maintenance that goes on in a very rigorous and regulated manner. The balance of the business plan also made a great deal of sense, since the planned new facility was also designed to relieve the commercial air traffic load in California.

The California market was getting so expensive that it made economic sense to have a cargo plane land at Buckeye, Arizona, unload the shipment onto trucks and then take Interstate 10 to California. Alternatively, the railroad was close by, and could be very competitive. I liked the project. We agreed on a commission and I went to work.

It takes time to conduct a review of any major financing, and one cannot approach a serious financial company without having done one's "due diligence"; you have to know the subject as well as the client does. If you make mistakes, there is no second chance.

I had no sooner told my wife that I thought we would be moving to Arizona than a call came from Larry Bell to say that the deal was in jeopardy. There had been a sudden increase in the price of land around the airport. The local owners had spotted a bonanza and wanted blood. Franck Curtin threw up his arms, closed up shop, and went back to France, and I lost touch with him. I shelved the business plan and began looking for another deal, one that I hoped would be more lucrative.

Yugo Riddles

It was not long before my wheeler-dealer juices were flowing once again. I am not sure how Drazen Premate came into my life, but this intriguing Yugoslav was in the business of raising capital based on an opportunity to launch one's experiment on the Space Shuttle. He was attempting to make absolutely pure cadmium telluride crystals in space, to use as "seeds" for a commercial exploitation of the technology on Mother Earth. So I had another flying project (albeit in space), which should have been relatively easy to fund based on the huge potential rewards. In our travels, Drazen introduced me to a number of professors at Embry-Riddle Aeronautical University in Daytona Beach, Florida and invited me on a trip to Yugoslavia to visit with some Yugo-entrepreneurs. What a delightful, pristine country, especially around one of the seven wonders of the world, Pritzvic. Too bad the place was blown to pieces in the war.

My time with Drazen was short-lived once it was discovered that he had allegedly used investor funds to finance a Yugo-rock band. He thought the music was great and he could make a bundle on the side. Unfortunately for him, few people agreed with his musical tastes and the rock band went the way of the Yugo car. The investors threatened a lawsuit and I ran away — but not before developing an excellent relationship with Embry-Riddle Aeronautical University.

I believe it was during 1990 that I was instrumental in creating what eventually became a public company, North American Technologies Group, Inc.. I had brought to NATK a technology to upgrade heavy crude oil into light and sulphur-free oil, which would have made Canada and Venezuela some of the largest oil producers in the world. This technology could also be used effectively for cleaning up toxic and hazardous wastes.

I approached the management of Embry-Riddle with the idea of starting an environmental school for cleaning up old airports and monitoring new ones. The idea was appealing, since the number of student pilots was down and the university needed new ways to raise working cash. I became the Director of Technology Transfer. I was invited to a number of board meetings where I met people from the airline industry, the FAA and other interested parties in the airline business.

I also met a number of politicians in Daytona Beach who

were always looking for new ways to bring business to the area. I decided to take another look at Curtin's Trans-Atlantic Aerospace deal, since McDonnell Douglas was shutting down its operation in Daytona Beach. I could not find the business plan, so I called Larry Bell and luckily he had one copy left. I sent it to Embry-Riddle for review. The professors at ERAU did not even respond, so I assumed it was of no interest.

Embry-Riddle now has an environmental school, as I understand, which will be invaluable as airports come under pressure to stop pollution. I eventually stopped working on the ERAU projects to pursue something more lucrative.

On My Own, Again

I had decided to form my own environmental company, called O.P.E. Environmental Industries, Inc.. I had acquired some new technologies and wanted to takeover Detox Industries, Inc., which was failing. I decided to call Tom Dardas and suffer more indignation, in the hopes of acquiring his company. Tom was his usual congenial self and invited me to take over the company; he even offered me an office from which to work. We were friends again, for the umpteenth time.

Tom is an unusual mixture of Mr. Charming and the vicious lawyer with a steel-trap mind. He is as quick with his wit as he is in turning a good situation bad and a bad situation good. Tom's parentage provided him with ample emotional genes, and his Mediterranean heritage shows, even though he has spent little time overseas — until recently. He possesses an unusual generosity with other people's money, but otherwise he has the lawyer's mentality, requiring payment for every waking hour. There was one occasion in which his genuine generosity would change both his life and mine, not to mention a few others. And that is another basis for this story.

I suppose it must have been at least 17 years ago that Tom met a very pleasant young Libyan student who was studying in the USA. His name was Adel Sennosi. This young man had been accused of embezzlement of some substantial funds and he was faced with a lawsuit, with no money. Tom took on the case *pro bono*, and won. Now, if you help a Libyan, as Tom had done, you will probably have him as a friend for life.

It must have been about the same time that I moved into Tom's office to take over Detox that Adel called Tom to see if he would be willing to help get the sanctions against his people, the Libyans, lifted. This was no small order, since anyone involved with the situation ran the risk of violating the terms of the sanctions, which applied to US companies and individuals as well. Tom had no intention of breaking any laws, especially given that, as a lawyer, he would be assumed to have known better. (What Tom did not know, at that time, was that I had been involved, very carefully, with an opportunity to sell Occidental Petroleum's Libyan operations to any company not subject to the sanctions.) We engaged the counsel of a local Houston-based attorney who, we knew, was well-connected in Washington and could guide us carefully through the legalities. Tom and I made a deal to work on the project together, and I felt reasonably comfortable knowing that I was working with two very competent attorneys. The object was to move very slowly and make sure that every decision was legal. Months went by as we laboriously gathered the data we would need to effect our deal.

And now, I must digress once more as I introduce another strand of the story.

To cut one long story very short, I am now a firm believer in Chinese acupuncture, herbs, and so forth, with an occasional Western drug — after it has been thoroughly checked for potential side effects. After our usual Tai Chi lesson, on Wednesday, July 17, 1996, Jan and I walked into the neighborhood diner.

We were waiting to be seated when I noticed that the TV at the doorway was tuned to CNN, announcing the crash of TWA 800. — A shiver ran down my spine. I knew what had happened. Section 41, the nose of the plane, had just broken off the main fuselage and fallen into the Atlantic, leaving the remaining passengers coasting on to an inevitable terrifying death from 13,000 feet.

Was this the same problem that had struck Air India 182 and Pan Am 103, and maybe others that had not been reported, or that I had missed reading about in the papers?

Raising Franck Curtin

I decided to find out as quickly as possible what happened.

First, I searched my office to see if I still had the Franck Curtin study lying in some file drawer; I searched in the attic. I could find no trace of the business plan. I called Larry Bell in Phoenix to see if he had a spare copy, but he had returned the last of his copies years ago. I called Dr. Steven Sliwa, president of

Embry Riddle Aeronautical University and Dr. Bill Motzel. Bill made a thorough search of their files but came up with nothing. I called Larry once more, to see if he had Franck Curtin's telephone number in Paris. He found it — but it had been disconnected. I then called directory inquires in Paris and many of the suburbs where Franck might be living. I was getting nowhere and feeling a little frustrated.

Then I had an idea. On July 21, 1996, I decided that the *New York Times* should be interested in the story, and they had correspondents in Paris who could conduct the necessary research to find Franck Curtin. I called the *New York Times* and spoke to Eric Smith. I briefly told Smith my story, and asked if he was interested in a follow-up. He was not. For him, I was just another jerk calling in with a wild story before the authorities had even had a chance to do any work on the case. (He is due for a severe reprimand, when his boss reads this book!)

I was at a loss, until I found my old address book, the one I was using when working on the Trans-Atlantic Aerospace project. Bingo! I had a number in Paris. But it was not correct. Was I going to have to spend a week in Paris, pounding the pavement? I was by no means reluctant, but how do you tell your wife you are dashing off to Paris for a mad adventure?

It must have been about 6:30AM when I was awakened by the office phone ringing in my ears. I normally let the voicemail pick up calls before 7:00AM, to discourage my zealous business associates who rise before 5:00AM and want me to go to work. This time I was interested to see if one of my contacts had any information. It was Franck Curtin, himself, saying that he had heard I was looking for him.

During my time at ERAU, I had visited Franck a second time round, at the elegant Beaumont Hotel, a 180 Swiss franc cab-ride from Geneva. Franck was clearly doing very well, but he was not interested in reviving his old project for Daytona Beach, Florida. We had a delightful dinner, and a great conversation, but he was clearly not interested in any of my projects. This time around he was quite different. He was very disturbed by my inferences concerning the state of the Boeing 747 fleet, which I had clearly drawn from his business plan.

He confirmed my suspicions, but warned me to just leave things the way they were. He warned me that the possibility of a major calamity was very real, that the prestige of several Western countries, especially the USA and Britain, was at stake and that the politicians and businessmen would play very dirty. I had better forget the whole episode, keep a low profile, and keep my mouth shut. If it were found out that I was investigating this problem, I might well be assassinated — and they would find me wherever I might be! Now those words are very serious. We only spoke for a few minutes but I got his phone and fax numbers, in case I wanted to talk to him again.

Now What?

Now I was caught in a dilemma. If I ran away from this explosive story, there was a risk of another Boeing 747 crash in which innocent people would be killed — maybe one of my own children — that I could have averted. Could I let myself be lured further in by the mystery? I mulled over the consequences for some days. I concluded that the story was not only a civic obligation but also, perhaps, a rare opportunity to make some serious money or fund a company with a worthwhile agenda. After all was said and done, I was now 58 years old, with no medical insurance and no pension plan; I was keenly aware that the social programs for spent entrepreneurs would be non-existent when my turn came to retire!

I mapped out a program of events and called Tom Dardas to a meeting at the Champs restaurant on the Katy Freeway. Dardas immediately recognized the potential of the story and how we might "parley" it into our deal with Libya deal (about which I have deliberately left you in the dark, for now). I told Dardas what I believed had really happened to Pan Am 103 and, indeed, TWA 800. This was real dynamite! I was now totally convinced that Libya was not the perpetrator of the alleged heinous crime of bombing Pan Am 103. This was something far more sinister. However, I needed to be quite sure of my facts, as I had no intention of emulating the British press and simply making up a story for the next day's sensational news.

Although no stranger to international acts of violence, the Libyans, if I was correct, were apparently innocent of this dastardly deed of bombing a jumbo jet full of holiday-makers. Could this have been a massive cover up?

Tom Dardas has a keen mind, especially when it involves the gentle art of removing money from one person's pocket and into his own. I supposed it must have been several weeks before Tom called me, full of excitement, to say that he had been summoned to Paris for a very important meeting. Maybe we were going to close our deal with on Libya, after all these months of waiting. I decided that he could use the opportunity to meet Franck and get a copy of the business plan. I introduced them by phone, and Tom invited Franck to dinner. The meetings took place in the Orly Hilton, near Paris. Whereas his meeting with the Libyans was probably a waste of time, Tom's meeting with Franck and his finance man was very fruitful.

Tom made a deal to buy the business plan (now known as the "Boeing Report") for $5,000, with a kicker if we succeeded. Tom gave Franck $500 in good faith money with the balance to come. It took us about a week to round up another $4,000, which we promptly wired to him with a promise for the balance very shortly. Franck then mailed to Dardas one section of the report — the chapter on "Section 41" in the Boeing 747's of early vintage. Section 41 is the nose of the plane to about the area where the stairs rise up to the upper deck.

This was just a small part of the total 570-odd pages. I was furious. Franck had promised the whole report, which would lend credibility to our analysis of the situation. We were at risk of appearing as though we had a mole inside Boeing, leaking us confidential information. If it looked like we had acquired the data illegally, the data probably would not be admissible in court. A knowledgeable, disgruntled ex-airplane designer could have fabricated this small part of the report. Worse, it could be construed as industrial espionage. We had to be certain the document trail was provable and admissible in court, as we would be faced with a formidable defense mounted by Boeing, the airlines and government agencies around the world.

Dardas seemed unconcerned about having received only part of the Boeing Report; he was confident that his agreement was valid and he did not need to reply on the Planegate fiasco to close the deal and get paid. I was frustrated, as I was spending a lot of time on the investigation with little or no control over when I might reap the benefits. I decided that we needed the report in its original form, and I was determined to make Curtin honor his agreement.

Dardas and I met in my home office on October 6, 1996 and called Curtin in Paris. I was determined to make this deal work, and I decided to tape the conversation. After a few of the usual pleasantries, we informed Curtin that we were taking notes; he agreed that that was OK. The conversation was 13.1 minutes long and it was very enlightening to say the least. That was it. I was confident of the outcome and decided to make a supreme effort to investigate this situation with all of my mental and cash resources. The mental resources were easy; the cash was not.

Chapter 2

"MY DINNER WITH FRANCK"

Evaporating Prospects

Where was I to go from here? I needed money, and the pickin's were thin. I had spent the last five years working on the North American Technologies project, seeking to save the Western world from the grip of the Middle Eastern oil cartel. I really believed the process of upgrading heavy oil to light oil was going to make me wealthy. Canada has over one trillion barrels of heavy oil in the Athabasca oil sands and Venezuela over two trillion barrels in the Orinoco belt. Compare this to Saudi Arabia, with only 200 or so billion barrels — albeit easier and cheaper to extract!

I was barely on speaking terms with my partners, over the apparent embezzlement of at least $1,400,000 that we should have received in March 1993. By virtue of my ownership of one third of the stock of a private company called Gold Spinners International, Inc., my two partners and I owned over 500,000 options on the Canadian company, North American Technologies, Inc., when it merged with a NASDAQ company to become NATGI (North American Technologies Group International, Inc.).

According to my partners, an international promoter named Robert Colgin Wilson had stolen the money. Boy, do I know how to pick them! Mr. Wilson, I learned in the February 24, 1997 issue of *Barron's*, was a 400-lb con man who had worked with such folks as Bert Lance and Chip Carter (President Carter's son) in numerous schemes. Well, I may have lost my money, but at least it wasn't in a two-bit con; these were the heavyweights of the business.

I went to Calgary, Toronto and Chicago trying to trace the stock. These guys were experts. The money had simply disappeared into an aluminum post office box in Toronto, courtesy of the Royal Bank of Canada, to whom I had entrusted the transaction by virtue of a "power of attorney." This event really sparked my interest in investigative reporting. It reminded me of the happy days in Libya, when I had genuinely enjoyed studying the behavior of oil and gas as they move in the underground reservoirs.

My investigation of the Boeing 747's became long and involved; I just hoped I was not wasting time, the way I did with this NATGI snafu. Briefly, in this case I called the Controller of the Currency in Canada, since a sizable amount of money had been stolen and removed from the country, which was probably a violation of the law. It was. The Controller told me to contact the Royal Canadian Mounted Police in Halifax, Nova Scotia. I was really pleased. I thought I was getting somewhere, since the "Mounties always get their man!" As it turns out, I was only one of hundreds whom Mr. Wilson had duped. He has been tried and found guilty of mail fraud and other charges in California and Florida. The fact that he would be languishing in a Miami jail (prisoner 40788-004 W/M BOB 4/1/41) until May 5, 2002 did not help my cause. I am still trying to find my money, but hope fades with each passing day.

Having become quite confident that I would never see a dime of the stock in Gold Spinners, I offered to sell my position for a small fraction of its value. My partners were very suspicious, and ran to their attorneys. They could not believe their ears at their good fortune, but I had better use for even a small amount of cash.

On the Road Again

I was now in a position to travel, so I bought a ticket to London, Paris Geneva, and elsewhere to secure my database. Was I crazy? I had been extremely active since the crash of TWA 800, clipping articles from the newspapers, making videos of the relevant news, researching in the bookstores and libraries. I had amassed over 100 lbs of materials.

Geneva is a delightful Swiss town of some 400,000 inhabitants, straddling the banks of the Rhone River alongside Lake Geneva. It was a very beautiful day as I checked into the Astoria Hotel. The Astoria is a cheap, at $100 per night, but like all Swiss hotels it is spotlessly clean and comfortable (even if you do have to hang your clothes on a hook in the doorway).

I enjoy any opportunity to immerse myself in the French culture. I am often amused to hear Americans raise their voices, with the mistaken impression that that causes their English to come out in French. Then there are the twitty Brits, who speak English with a phony French accent — with equally poor results. This all amuses the Genevois, who invariably speak beautiful English, in a somewhat condescending manner.

The next day, I called Franck Curtin in Paris and arranged to have dinner with him and his new wife, Janine, the following evening. Meanwhile, I was very nervous about my suitcase full of extremely important, extremely heavy documents. It was going to be a major pain to haul them to Lockerbie and beyond. I set out to visit the railway station in Geneva, to see if there were any lock boxes. There were — but one could only leave one's luggage for 24 hours. The porter explained that it was a security measure; they were worried about terrorists. On further discussion, I determined that after two days, they would remove the luggage and store it elsewhere in the station. I knew I was going to be gone at least a week, so I left the case unlocked, with a number of business plans on solar energy, in English, on the top. I knew those documents would not interest the porters. The fact that I left my case unlocked might have given them confidence that there was nothing dangerous in it. I took the chance: I decided the odds were greater that someone might stop me at Customs. I was a little paranoid, I must admit.

Late that afternoon, on Saturday November 2, 1996, I took the "train à grande vitesse", or TGV — the train that goes 180 mph. Quite an experience. They are so fast, the ride is so smooth, and the experience is more comfortable than the airlines. I arrived at the Gare de Lyon, south of Paris, and took the RER metro train to Orly Airport in the north. I checked into the Orly Hilton Hotel, and called Franck.

At about 8:00 PM, Franck and Janine turned up in a car designed for one small French lady. Janine was very apologetic, as we forced our bodies into every corner of the car like plastic in an injection mold. I really did not care. I gladly would have ridden on the roof, I so desperately wanted to talk to Franck about the Boeing situation. I had not lived in Paris since 1977, when everyone still dressed up for dinner on a Saturday night. Franck and Janine had clearly adapted to the American way of going into any restaurant in a T-shirt and jeans. I was seriously overdressed, and my pinstriped suit was developing fine pleats thanks to this car ride.

The restaurant, Chez Margot, was crowded with steak tartar-munching, Grand Cru-drinking, Gitane-smoking French, clearly specimens living high up the food chain. I loved being back in Paris, even though life in America, under the environmentalists, had trained me out of my tolerance for smoky restaurants. The conversation was intense. Janine disliked the smoky atmosphere, too, but she was very pleased to be out and listening to the stories. We spoke French all evening (which pleased me intensely, since I thought we might attract less attention from our neighbors, who were very close indeed). The rules of the game of dinner are entirely different in France. First, space is as tight as it is in the cars. One can easily be placed only a few inches from the next guy, whose dog may well stick its nose into your crotch. On the other hand, time is plentiful; you have all the time in the world to eat, drink and converse — which is exactly the opposite of the approach in the US. In the US, an overly friendly waiter will greet you, and then insist that you converse with him, follow his advice, and obey his demands. He will bring your meal in the minimum of time, then spend the next ten minutes trying to get the plate back. Here in Paris there was no "My name is Pierre, I will be your server tonight!" And the atmosphere was electric.

I immediately thought of the movie, *My Dinner with André*, in which two individuals explore a variety of subjects over dinner, in the most relaxed fashion, and with a certain delight for conversation. This was, "My Dinner with Franck" and Janine's presence only added to the delight. We must have talked for half an hour about our lives and how we looked on the world, before the conversation got down to the subject and reason for the meeting. The wine was flowing freely and the tongues were well oiled.

Now, I realized I had to be less flippant and arrogant with my remarks, because the conversation turned very serious. We talked for hours about the Boeing situation.

Frank Talk about Boeing 747's

Franck spoke very clearly and deliberately, as he described the problems with the early vintage Boeing 747's. He told me that the problems were caused by Section 41 separating from the main body of the plane, usually in-flight, although there had been cases where the nose of the plane had collapsed on the ground while taxi-ing. He total me that Boeing had informed him that seven planes had crashed with the same problem. Those loaded with passengers, of course, got big media coverage — but if a cargo plane goes down with three pilots, it rarely gets a mention. You will recall the photographs of the Pan Am 103's nose section lying in a field in Tundergarten, some four and a half miles from Lockerbie; and both TWA 800 and Air India 182 suffered a similar situation, with the nose being found many miles behind the rest of the wreckage.

Franck took out his pen, and began to describe what he had been told (during the meetings with Boeing and the FAA) would be required to fix these planes. This was during the time when he was striving to manage his part of the massive repair program.

The Section 41 is assembled at the enormous Boeing facility in Everrett, Washington. This Section costs about 40% of the total cost of the plane. It contains all of the control functions and monitoring functions; it is a very complex piece of engineering. Once this section has been completed, it is winched over to another part of the plant where it is joined to Section 42, the main fuselage. Boeing aircraft parts are manufactured all over the world and are brought to Boeing's plant for final assembly. As Franck pointed out, it is like a giant Lego set. I would like to point out that the nose sections were also built separately in the manufacturer of the Comets in the 1950's. It is clearly the most cost-effective way to manufacture large commercial planes and technically is quite sound. That was not the problem.

The first Boeing 747-100 rolled off the line in February 1969.

At 11:34 AM on February 9, 1969, Boeing 747 RA001 (N7470) launched into the wintry Washington sky for its first test flight. Interestingly enough, the maiden voyage was abruptly cut short when an F-86 chase plane noticed the inboard end of the central flap had come loose. The crew decided to play safe and returned to Paine Field.

There was no question this plane was a major breakthrough in airplane manufacturing technologies; and it has turned out to be a very significant contribution to the world's commercial aviation business. Over 2500 Boeing 747's have been built and they ply the world continuously with passengers and freight. The early 1970's were not easy times for many businesses, such as airplane manufacturers, and Boeing was no exception. In fact, you may recall that one wag suggested that the last person to leave Seattle should turn out the lights. Therein lies the problem.

Franck leaned back in his chair, picked up the glass of red wine, sniffed the contents and, pleased with his choice for the evening, paused while the fruity liquid was being analyzed and compared to previously tasted wines of years gone by. He was relaxed, while Janine was beginning to show signs of not enjoying the densely smoky room (knowing full well that the night was far from over).

In the early 1970's, Boeing had a problem. The aluminum it required for the 747 was a special grade, which had been carefully tested by the structural engineers to withstand tens of thousands of take-offs and landings. More important was the life span of materials, which would be subjected to hundreds of different stresses and strains as a plane flew through turbulent weather, with its cabin pressurized to about the same pressure it would experience at 7,000 feet. The pressurization is necessary for creature comfort and indeed for life support at higher altitudes. The cost of the aluminum was such that its profit margin could be made acceptable to the shareholders only if Boeing could acquire the aluminum from someone who could produce it at a lower price. According to Franck, Boeing found one such manufacture in the old USSR, now Russia.

The first ten planes, he said, were built with the correct grade of aluminum; but then the Russians started taking short cuts. Did the Russians do this on purpose? After all, a Cold War was being waged and a problem with the Boeing 747 might simply be an expression of Western decadence. Boeing carried on making the planes at an ever-increasing rate. The first planes were the 747-100, which could carry 100 people; then came the 200 series and the 300 series, until Boeing had built 686 planes.

Something then triggered Boeing to verify the quality of the aluminum and it discovered that the materials in these planes were substandard. There was a moment of silence when I considered how many of these planes I flown around the world with my family. Had I been on one of these planes that had subsequently crashed? What if another one goes down with 300 people? I had known these facts from the famous recorded telephone call, but now, to hear it straight from Franck's lips, while observing his other body language, was very disturbing.

I changed the subject to ease the tension and to bring Janine back into the conversation; she had been ignored in an ungentlemanly manner for the past thirty minutes. The waiter came by with the dessert tray and Franck suggested I try the "Belle Margot", while Janine settled for some assorted ice creams. Then we had coffee. I was not ready to quit; I had not traveled half way across the world to give Curtin the balance of the $5,000 without getting all of the information I could. Janine would just have to wait. Franck was relaxed and in no hurry to leave.

We ordered some Napoleon brandy, and I posed the next question. What do you know about Korean Airlines 007? Did it suffer the same fate as Pan Am 103 and TWA 800? Franck pulled out his ballpoint pen and began to tell yet another fantastic tale. Franck clearly liked to talk with a pen in his hand and I certainly did not discourage this method of communication, since I had every intention of taking the napkin with me.

Franck began by explaining that US Intelligence was particularly interested in the southern part of the Island of Sakhalin, where the Russians had a major naval base and highly advanced radar monitoring systems. The Korean Air Lines flight 007 was on a flight from New York to Seoul, Korea, via Anchorage. In the early hours of September 1, 1983 the plane completely disappeared with all 269 passengers and crew on board. According to the official reports neither the plane nor any passengers were ever found. You must remember that these were early Ronald Reagan years and Ronny was a major anti-Communist declaring that the Soviet Union was an "evil empire." What happened on September 1, 1983, exactly 14 years after Colonel Gadhaffi took over Libya, could have been the beginning of World War III.

Tensions were very high and the CIA was alert to the potential conflicts, as detailed in Michel Brun's book, *Incident at Sakha-*

lin, *The True Mission of KAL Flight 007*. What happened to Michel Brun, once his book had been published, is another major story; but we have our own story about the investigation of KAL 007 and what Franck understood had actually happened, based on his discussions at Boeing.

Mr. Curtin had been working very closely with Boeing's management to prepare for the needed repairs to the Boeing 747's. Drawing on the napkin, Franck went on with great authority, to demonstrate that the CIA had modified the Boeing 747 operated by Korean Airlines; and that it had been over-flying the area off the island of Sakhalin every day for many weeks — each time getting closer, by six nautical miles. Each time the Korean Airliner violated the Russian air space, the Russian authorities would contact the US Ambassador and lodge a very strongly-worded complaint. The US Air Force was having a field day, since each time the civilian Korean plane came close, the Russian radars would light up and the Americans would gather important military information. All countries spy on each other so there was nothing new with this process. Soon, the Korean airliner was actually passing over the island itself, and the Russians made it very clear that these flagrant violations would result in the plane being shot down, as it was clearly on a military mission posing as a civilian airliner.

Who was playing these dangerous games? Both parties? I have read nothing more than you may, yourself have read as to the Russian point of view; but I have heard firsthand what Franck learned had happened.

The brandy was definitely migrating to the brain. I was beginning to feel like James Bond on a mission, but without the glamorous girl. Franck explained that Boeing had spent millions of dollars re-designing the front steering wheel-well to accommodate sophisticated cameras. The idea was that the front wheel could be lowered, during flight, without the main wheel coming down, thereby exposing the camera to the ground below. Ingenious, I thought, as I visualized the co-pilot *cum* cameraman mapping the island of Sakhalin with these sophisticated instruments.

Then came the big one. Franck told me that the CIA had paid the Russian and the Korean governments $100 million dollars *each* in compensation (or hush money, whichever term is appropriate).

Janine was relieved to see that this was the final episode for the evening; and we rose to take our leave. As we left the table, I decided to ask Franck if I could keep the napkin. I was concerned that he might be offended if I just pocketed it. "Ça vaut pas la peine!" he whispered to me, as he took Janine's arm. "Whatever", might be the modern version of the translation. I picked up the bill, and the napkin, and was very satisfied with a delightful evening and the underlying conversation. The air was quite chilly as we attempted to force the bloated torsos into the miniature vehicle on our way back to the Orly Hilton. I arranged to meet Franck in the morning to give him the remaining $500 for the Boeing Report, and I expected to be on my way.

My mind was racing, and I could not sleep — for three reasons. One, I was too bloated with French cuisine, two, I was jet-lagged and three, I was too occupied with my memories of that fateful flight of KAL 007. I looked up at the ceiling and my mind raced back to my thoughts on September 1, 1983. As a pilot, I always take a particular interest in crashes; I had been trained to read accident reports, so as to learn from experiences I hoped never to have, myself. Furthermore, September 1st of any year brings back vivid memories of the Libyan coup.

You will recall that I had flown with Bondy from Perth, Australia to Bombay in the cockpit of a Qantas Boeing 747 and had been particularly interested in the inertial navigation system. Commercial airlines only take enough fuel to stay in the air for one hour after reaching their destination; so it would be recklessly foolish to wander off course. Those Korean pilots were too experienced to be off their intended course.

Furthermore, had these Korean pilots been "asleep at the wheel" and had they maintained course in that "erroneous" direction, they would have ended up in North Korea or China, where they would have been shot down anyway! I do not believe a word of the official version, any more than I believe the official versions of Pan Am 103, TWA 800, Air India 182 and others.

Then I remembered that I had been upset by the newspaper reporters and even more so with the naïve TV talking heads, who expounded their theories with no knowledge of trans-world flying. Not one reporter even bothered to bring a Boeing 747 pilot to the studio for an interview. I was so upset at the time that I recollected having met a Boeing 747 pilot on a business deal some time earlier; I searched my files, and Bingo! there was his card. I called Frank Branham in Marrietta, Georgia and tried to get him to give his opinion. He was clearly put out at the prospect of being on national TV, so I gave up.

I found it remarkable that a plane as large as a Boeing 747, with 269 passengers, would simply disappear. The Sea of Japan is not so deep that a search would not have found at least one major part of the plane. Could it be that a major effort to find the plane did in fact take place — to hide the fact that the nose wheel had been modified to include a camera? Too many questions were unanswered. Did the CIA pay the Russians and the Koreans $100 million each, and we do not know about it? We know that the CIA's budget exceeds $26 billion *now*, and we are not threatened by a Cold War. $200 million would be peanuts to the CIA, and the information was worth more than that.

I felt a cold shiver run down my back at the thought that yet another Boeing 747 could disintegrate in mid-air, with the loss of another 350 people — some of whom I might know. What would happen if I published my findings on the Internet, at this early stage, before I had completed my investigation? If 600 or more planes were grounded, it would certainly bring a major part of the world's commerce to a grinding halt. Was I becoming an accomplice to the cover up, or was I simply out of my depth in such a huge undertaking?

What was the motive behind most of the wars on this Earth? Usually a simple case of economics, power struggles and religious intolerance. Was this any different? Finally, since I was a professional engineer, a pilot and quite capable of careful analysis of technical problems, I decided that needed to be sure of my facts. Exhausted, I decided I was right to be cautious and fell into a deep sleep.

My "Breakfast with Franck" began promptly at 8:00 AM as

planned. We chatted for a while over the *café au lait* and a couple of croissants, and I could see that Franck was anxious to see the money. I was waiting to see the Boeing Report. Frustrated, I pulled out five $100 bills, handed it over and asked for the report. He had not brought it. I was furious when he said the copies were all in an office in Nevada and he would have one Fedex'd to my office. What could I do, make him mad and get nothing? I decided that the agreement had been signed with my partner, Tom Dardas, and I would have to have Tom put pressure on Curtin to honor the agreement. I was beginning to think that Franck was now concerned that he had said too much over dinner. That is not the end of the story, but it will probably be my last dinner with Franck.

I checked out of the Orly Hilton hotel and took the RER to the Chunnel train. I was on my way to Lockerbie.

Chapter 3

LIBYAGATE

Tunis. Back in 1972 when I bought that MGB sports car, my wife and children were spending the summer in the UK, and hence I was free to do as I pleased on the weekends. I decided to drive my new car to Tunis, via Djerba. Djerba is an island of sand in the southwestern corner of the Gulf of Sirte and gives new meaning to the idea of an oasis. On this island are a number of hotels of modest quality, catering to sun-worshipers from Northern Germany and Scandinavia. On this island-paradise, anything goes, and the rules of normal conduct anywhere in the Arab world simply do not apply. Furthermore, the offshore oil from the area caused a rift between Libya and Tunis over ownership; I believe, in the end, the Tunisians won the dispute.

I took the coast road towards Tunis and passed through Sfax, the date capital of the world, checking out a few Roman ruins along the way. The Tunisians speak fluent French as well as Arabic. North of Tunis is a delightful Arab village called Sidi

Bousiad, where I thought I might retire one day; would they let me? The quaint villas all have heavy wooden doors painted a pleasant light blue; the nail heads are painted black.

Why was Thomas A. Dardas, President of Detox Industries, going to Tunis on December 12, 1994? He had spent over $12 million in an abortive attempt to cause bacteria developed at the University of Houston to eat the highly toxic waste called PCB (polychlorinated biphenols). I could not imagine there was a toxic waste problem; there was no great market opportunity for dates, and the oil concessions were owned by the state and the major oil companies. So what was the purpose of the visit?

On January 27, 1995, Dardas had told me the story of how he had helped a bright young Libyan to resolve a serious legal problem, *pro bono*. During my five years in Libya, I was impressed by the strength of the family ties and the warm relationships between friends. Dardas continued with an interesting story of how his friend, Adel Ali Sennosi, had received US refugee status and was then living in Denver, Colorado, working as a real estate agent. Adel's family remained in Libya and he was all too aware of the hardships caused to the Libyan people by the sanctions. He was anxious to see if he could find a solution, and have the sanctions lifted. I knew from other sources that Libya had tried many times to find a political solution to the crisis, which was having a serious effect on the daily life of ordinary citizens. Each time, I was told, the high-ranking foreign politicians who were offering to help had demanded large sums of money, millions of dollars. None had come close to being successful but, naturally, they kept the money.

In early 1995, the United States was in a political frenzy. Clinton had trounced President Bush and became president in 1992; now the Republicans were anxious to take back the White House. Clinton was equally determined to stay, and knew that the way to win was to raise large sums of soft money. Irregularities of the Clinton/Gore fundraising activities came to light, such as the Lincoln bedroom becoming a motel for major donors.

Around the same time that Tom was trying to help Adel Ali Sennosi, I was attempting to sell the Oxy operations in Libya to an Indian oil company (who were not affected by the ban on international relations with Libya). I was well aware that the sanctions applied to American oil companies and, when they were imposed, Oxy signed a "Stand Still" agreement with the Libyan National Oil Company. Under this agreement, the revenues from the oil sold from the Oxy concessions was to be placed in an escrow account. These funds would then be available to Oxy the day that normal diplomatic relations were re-instated between Libya and the USA. At the time we were involved, the cash in the escrow account was over $500 million dollars! Oxy was confident it would never see a dime of it, and was willing to sell the concession to any non-US oil company. My deal did not consummate — the price demanded was too high.

I thought it made sense to hire the same Houston lawyer who was vetting that deal to work on any transaction that may involve Libya and the Dardas deal, as well. On April 17, 1995, I introduced Dardas to Pat Strong at his office in Houston, Texas. On May 2, 1995, Dardas went to Tunis again for more discussions. On June 8, 1995, Dardas was due to go to the White House for a "photo-op" with President Clinton but, I believe, the contribution to the DNC was higher than the funds available to him at that time. On June 9, Dardas met with the FBI in New York, on June 13-15, he was in Washington at the State Department then back to New York with the FBI on June 19, 1995.

Dardas is a brilliant negotiator with a quick mind. He had done his homework when he met with the Libyans and when they asked him what it would cost he said, "Nothing! — but if we are successful, then $1 billion!" Did I say billion? Dardas chuckled when he told me the amount. "Those kind of numbers make people believe you are for real." The fact that he demanded nothing up front was a stroke of genius, and curiously enough it is the basis of many deals in the US. Perform, and we'll pay. The deal was only verbal, at the time, but you would have thought it was signed, sealed and delivered to the Washington courthouse, he spoke so confidently about the prospects.

The FBI had the two Libyan suspects, who allegedly bombed Pan Am 103 over Lockerbie, on its Ten Most Wanted list. There was a $4 million bounty for anyone who could deliver them to FBI custody. Under the Dardas deal, the two Libyans would be sent to Budapest, where the FBI would capture them and take them to Washington, DC. I am now glad it all failed, as my instincts tell me that the two might have ended up being killed for "resisting arrest". I'll explain later.

Dardas was commuting from Houston to Dallas, to New York to Washington on a regular basis. When Dardas believes he can close a deal, he will work night and day, travel halfway around the world for a one-hour meeting and all this despite a lung problem, which often left him unable to speak. These were exciting times. We were quite certain that all calls were being monitored so Dardas would keep saying, "as long as it's legal," during any conversation. Dardas had offered me one eighth of the deal for bringing in the Oxy deal and for advice on how to deal with the Arabs.

The American frontal-attack way of doing business would get you nowhere in Libya or any other Arab country. The Arabs have been desert traders for thousands of years and they know the best deals are made if you take your time — lots of time — to the point of distraction! For a Greek temperament like Dardas, this was a hard lesson to learn, but he was determined to complete the project and was prepared to sit in hotel rooms in Tunis and elsewhere for days on end, waiting for a meeting. I told Dardas how Armand Hammer had won the Oxy concession in Libya — by patiently waiting for weeks in the Waddan Hotel in Tripoli. Now, there was lots of activity, getting all the papers in order; and Dardas was busy raising money for the effort by selling a participation in the deal.

Then came the big bombshell. On June 27, 1995, word came down that Clinton wanted half the money. What! I thought "Libyagate" was in the making.

I'm not sure who passed this message to Dardas, but I believe it was Roger Clinton, the President's half-brother. It could have been the FBI; but I doubt it. Another person who appeared to be involved, as the designated bagman, was Anthony Lake — the would-be Director of the CIA. Lake, a friend of Clinton and his former National Security Advisor, was sworn in before the

US Senate on March 12, 1997, to confirm this position. Then, very abruptly, about March 19, Lake had withdrawn his candidacy in disgust. I wonder what the Senators really knew and why they were against him running the CIA?

$500,000,000?! I thought this guy was a socialist, concerned about the poor! At least we were to invest the money in worthwhile projects.

Dardas seemed pleased with the deal to share the funds with the President, and I was sure we would see nothing at all. I was confident that the deal would quickly go behind closed doors and we would be out in the cold. Looking back, it probably already was being handled by a "Back Channel." Furthermore, Gadhaffi is no lightweight politician and would look on this payment with a very jaundiced eye. If the word got out, the Clinton administration would accuse the Libyan government of attempted bribery and would take political advantage of the incident. Dardas was determined to go ahead, but I believed that once Gadhaffi knew Clinton wanted $500 million, the deal was dead.

The story, however, goes on. The Libyans were naturally skeptical of Dardas's ability to deliver, that is, to get the sanctions permanently removed, and wanted a photograph of Dardas with Clinton. Dardas made the approach and the price was now $65,000. Dardas was also paying someone with connections some $50,000 per month, or at least that is what he told me in a money-raising meeting. He did not have $65,000 dollars, so that would have to wait.

On August 3, 1995 Dardas went to Tunis and on August 4,

1995 he signed the agreement with the Libyans. Now the money raising became much easier, and a minimum of $525,000 was raised. It could have been as high as $850,000 for the project. Dardas was on a high.

Then, in early October 1995, a letter came from David H. Harmon, Acting Supervisor, Enforcement Office of Foreign Assets Control, Department of the Treasury. It was a "Cease and Desist" order with respect to the Government of Libya. Dardas said it was a "cover yer ass" letter, in case anything went wrong. He carried on with the project. It was, most certainly, a "cover yer ass" letter — the process went on, unimpeded. On November 10, Dardas announced that he was going to Budapest and the deal was going forward; instead he went to Tunis.

Then, on December 7, he came back from Europe and said that the "capture" of the two Libyans would be announced on Larry King Live, the nightly CNN talk show. Nothing happened for months.

Clinton had beaten Dole, hands down, and in some ways it was a good thing; Dole was left with advertising Viagra! No real balls! Dardas was very comfortable continuing the project, and on January 11, 1996, made another attempt to "buy" a photograph with the President; the price had risen to $100,000. Dardas declined, as the money could be better employed on the project. More months went by, an occasional call to Libya when he could get through. He knew that all of these calls were being monitored and made no bones about it. At one meeting with the FBI they showed Dardas photos of him in Tunis — he was being watched on a daily basis, wherever he was.

Then a very interesting development occurred, which I took as positive. On June 23, 1996, there was an Arab Summit meeting in Cairo and Christiane Amnapour interviewed Colonel Gadhaffi, which was shown on CNN. I made a video copy to show Dardas. It was more than clear that Gadhaffi was reaching out to Clinton and the American public. Dardas didn't care; he was convinced the deal was going to work anyway. One problem he had was a constant shortage of money. On July 12, we both went to Denver to meet some potential investors — they were not interested but I did meet Adel at the Damascus restaurant in Denver. Adel was very nervous that I might be an agent, but Dardas assured him that I was not; we had a pleasant chat and an Arab lunch.

Then came another startling and unfortunate occurrence, which curiously enough would contribute to our effort. On July 17, 1996, TWA 800 plunged into the Atlantic, just offshore from Long Island.

What Franck Curtin had told me, I related to Dardas on August 12; five days later, Dardas was on his way to Paris to meet another Libyan and to make the deal with Curtin for the rest of the information. The Libyan meeting was fruitless, but the Curtin deal was well worth the trip: Dardas gave him $500 down, on the promise of receiving a full copy of the report, and came home to Texas. On October 6, we got together and called Curtin in Paris to discuss the data, which he had not yet delivered. We told him we were taking notes, and since he clearly did not object, we recorded the conversation; it was 13.1 minutes in length. 13.1 minutes of dynamite.

Up until this point in time, I had merely been fascinated by the Dardas efforts, but I was very skeptical about the whole situation. I thought it was unwise to fool around with the Clinton Administration, as Clinton cared little for his oath of office and was abrogating the US Constitution whenever it conflicted with his own political agenda. The economy was growing and the American public took little notice of Washington affairs. The liberal-biased news media was pumping out poll after poll in Clinton's favor. The Republican-dominated Congress just stood by and watched. Clinton was amazingly charismatic, a great actor. (There is a telling video of Clinton and another fellow walking back from the funeral of former Secretary of Commerce Ron Brown. They were sharing a jovial moment, when Clinton spotted a camera — he immediately stopped smiling and "wiped the tears" from his eyes.)

The American public was making money. They were no longer interested in cleaning up toxic waste, since it had become clear that it would cost them a bundle; so our "Detox" business was collapsing. I decided to follow up with an investigation of the Boeing 747 crashes and left Houston on October 31, 1996, for Geneva, Paris for my "Dinner with Franck" and onto Lockerbie as you already know.

Time just kept rolling by and still Dardas believed the deal would go down if he were just patient enough. For a piece of one billion dollars he was going to be patient and anyway the investors were getting concerned about their substantial participation. On February 8-9 Adel came to Houston and stayed with Dardas. I was invited to come over, as it was Adel's birthday. I told Dardas I had some very interesting supper 8 movies of our time in Libya, they were well done and had a minimum of family included. I went to Dardas's house and met Adel for the second time. He was extremely nervous. When I shook his hand it was weak and sweaty a sure sign of stress. I showed my movies of Tripoli, Leptis Magna, Sabartha and a general look around town. Adel showed little enthusiasm for the movies, which I noticed, had a huge 20-30 foot poster of King Idris hanging from the outer wall of the Tripoli souk. It was clearly pre-Gadhaffi days. King Idris was of the Senoussi tribe but Adel showed little interest in the poster so I assumed he was of another Senoussi family. It was becoming quite clear that all was not going as planned. Yet on June 23, 1996 Dardas visited with the FBI in Dallas to find out that the agents were getting very upset with the lack of progress. Again he seemed positive and on June 24, he asked me to line up a private jet to take these two Libyan suspects to Budapest and then onto Washington. I called around and found that the price was going to be about $70-80,000, which we did not have. Then all hell broke loose. Another investor in Detox bought a satellite phone so we could communicate with the folks in Libya. Dardas raced to the airport to meet with the folks at the Abus Nawas hotel in Tunis. I went to London to be of help if need be. Then again nothing happened.

Months of phoning and traveling and going no place. On November 6, 1996 Dardas left for Tunis once more; he was so ill he could hardly talk. By the end of the year Dardas, after a gigantic, relentless but pathetically useless effort, handed me the Boeing Report and bowed out of the project. I decided I would approach the project a different way, making sure I was operating legally and you are now enjoying reading about my adventure.

For those interested in looking into Libyagate there are many documents in the Dardas files, phone records and travel itineraries and you might be amazed who's names will appear. The investors in the project ate their losses and had no recourse since their greed had them in a no win position. They could have sued Dardas, and you can imagine the field day the lawyers would have with this case. The "Ragin' Cajun" would have gone ballistic in defense of the president and the FBI would be hiding under Janet Reno skirts to avoid yet another independent counsel! But that was then, so now, with Clinton known as a consummate liar, the facts can come out and chips can fall where they may!

There is so much to tell but curiously enough there were a few stories I should briefly bring to your attention.

On January 26, 1996, John Lancaster of the *Washington Post* had an article entitled, "Gadhafi and Farrakhan to Sway US, Libyan Says." It turns out that Louis Farrakhan, leader of the Nation of Islam in the USA, had visited Gadhafi the previous Tuesday and pledged one billion dollars to "Muslim causes" in the USA. WOW! *This was exactly the same amount as the Dardas' proposal!* By March 15, 1996, Farrakhan had come under considerable attack for his dealings with Libya; and he publicly claimed, "I'm no foreign agent." By August 27, 1996, a *Washington Post* article said "Farrakhan wants his $1 billion gift." Then came *Time* magazine, September 9, saying "A Fool and His Money". I quote, "What a pity that the Clinton Administration won't let Louis Farrakhan accept either the $1 billion gift or the $250,000 that went along with the human-rights award he got from Libyan strongman Muammar Gadhafi last week."

Several high-ranking Libyans were involved with the negotiations with Dardas, but Mr. Ibrahim al Bisarhi was of particular significance; he was a major player in Libyan intelligence operations. I was surprised to read that he was killed in a car accident, at about this time. Most unfortunate for him, but also unfortunate for Tom Dardas, and very disturbing for everyone involved.

Chapter 4

LOCKERBIE, SCOTLAND. PAN AM 103 INVESTIGATION

The Chunnel Train beats flying to England and it beats the ferries, hands down. The system for boarding in Paris is clean, efficient and well organized and one quickly finds oneself racing at 300 kph. The trip under the English Channel (or *La Manche*, if you are French) is so quick that you might easily miss it, if your reading material is at all engrossing. The most amazing aspect of the trip is that the speed drops from 300 kph to 30 kph on the British side. The Brits simply did not get around to replacing the rails that were probably laid by Mr. James Watt in the 18th century. Many of the Brits were opposed to the Chunnel, and without everyone fully "onboard", it is a wonder that the project could be completed, at all. When the French and British met somewhere under the Channel, the Brits knew they had met the Frogs by the smell of garlic! England is such a quaint place.

I rented a car at the Waterloo Station and quickly found myself heading north to see my folks, in one of the nicest, oldest market towns in England, Swaffham. Swaffham Market was the place where the Brits invented bartering for goods and services, and now it has its own power generating windmill. It was a time to discuss the Comet crashes with my Dad (and to get the clothes cleaned at the laundry, and sample mother's green tomato chutney in a cheese sandwich; you won't find *that* in Paris!). I did not tell my folks what I was up to — they already think I'm crazy.

Traveling north and south in the UK is a snap, since the motorways are fast and effective. Going from east to west is a different matter. I was completely exhausted from changing gears and yanking the steering wheel to avoid the various obstacles the crazy Brits place in the road. Apart from the roundabouts (which work very well if the traffic is light, and not at all if its heavy), they frequently block off half the road, on one side and then the other. This gives to the uninitiated the terrifying experience of heading straight for the oncoming traffic, with no idea who has the right-of-way. I think they call them bollards.

As I mentioned earlier, when one thinks everything is going wrong, it quite often turns out to be alright. I stopped at a very nice, American-style motel in northern Lancastershire, and was amazed by the appearance of the people. They all looked like me (though with a lot more hair). Then I realized that my father was from Lancastershire, and that explains it. I must say, they all looked very healthy, with clear ruddy cheeks. Maybe I should never have left my roots; I can certainly mimic the local dialect. The Brits are a very warm and well-mannered tribe, which pleases me each time I return (even though I could not wait to emigrate in 1961).

The next morning, the wind was blowing "like billy-o" and my little car was hard to control as I raced toward Lockerbie. I made it, nonetheless, and found it quite as I expected: a small market town full of Scots. I drove around the town for about thirty minutes, and there was absolutely no sign of the tragedy, not even a memorial in the center of town. I was beginning to feel I had wasted my time.

It was a beautiful day but the icy wind could cut you in two. I spotted a local coffee shop, which seemed to be the only social center in the entire town. I opened the door and a dozens eyes turned toward me. The room went quiet. The manager broke through my discomfort, saying, "Cum on in, luv, whort kin I git yer?" The place was full of Scottish women, each with a bonny baby; they soon went back to gossiping and smoking. A pot of hot tea and a cheese roll soon gave me the courage to look further.

I drove around again and was about to give up, when I spotted the local library, which had been obscured by the scaffolding on the pub next door. I parked, and asked the librarian if they had anything on the Lockerbie crash. I had hit the jackpot. There was a whole wall of books and other materials, so I settled down to conduct my study. At first, I thought the two librarians were too curious of my activity. About every thirty minutes, I had to ask for change to operate the copying machine and I could tell they were keen to know what I was working on. I told them someone I knew had died in the crash, and I was making a book. That seemed to satisfy their curiosity. There were books of photographs on a table, showing where the memorials had been constructed. I asked for driving directions and took my leave, with my bundles of copies of data. It was quite clear that others had been before me and were also questioning the official version of events. With my aviation sources, I had unique information.

The Dryfesdale cemetery was only a few miles down the low road to Dumphries. The trip to the cemetery was through some of the most beautiful, pastoral winter countryside that anyone can imagine. The fields and forests were a mass of different shades of green with the early morning dew shimmering in bright sunshine. The cows and sheep were classically content, as they wandered about on the rolling hills. It was impossible to visualize such a ghastly scene as a 747 falling out of the sky into such a peaceful scene.

As I approached the cemetery, I was stopped by a road crew resurfacing the country lane; they asked me to wait while the trucks were unloaded. I decided to park the car and walk the rest of the way. I entered the gates to the Dryfesdale Cemetery and could see no sign of the memorial. Still, I thought, it must be there; so I began to walk to the far end of the property. Judging by the inscriptions on the headstones, this was a very old cemetery, indeed. Then I saw it. To the right of the walkway and along an ancient rock wall was the memorial to the victims of the crash of Pan Am 103.

A shiver ran down my spine. So many people! Two hundred and seventy innocent victims, with ages ranging from eight months to eighty years, their names engraved on a marble slab. It was not an elaborate memorial, but very dignified. I took several photographs, as I slowly walked towards the stone wall where several plaques had been mounted. I began to read them, and one inscription in particular caught my attention:

> ALEXIA KATHRYN TSAIRIS
>
> July 6, 1968   December 21, 1988
> INNOCENT VICTIM OF TERRORISM
> "They never die, who have the future in them."
> THE ALEXIA FOUNDATION FOR WORLD PEACE
> Franklin Lakes, New Jersey, USA

I choked. Then I sobbed out loud. This young lady had been the same age as my children. I was completely overcome by emotion, when a hand was placed on my shoulder. I had not noticed one of the workman following me to the memorial. He offered to take a photograph, which I declined, since it did not seem appropriate at the time. I slowly wandered back to the car swearing to myself that I would expose this story, one way or the other. I just had to. I had an obligation to make sure nothing like this happened again. But I was playing a dangerous game.

It must have been an hour before I had regained sufficient composure to drive to the place where the Pan Am 103's Section 41 had landed, in a field near Tundergarth. I drove back through Lockerbie, along the narrow streets in town, then up the hill towards Tundergarth. Again, this picturesque village looked at peace as it lay bathed in the bright winter sunshine. Suddenly, at the top of a hill, I saw what has to be one of the oldest country churches in Scotland. It is very simple edifice, nestled near a forest in the otherwise farming area. I parked next to the church and walked over to a small building that had been converted to another memorial to the victims of the Pan Am 103. The notice on the wooden door told me that the key was in the little cottage next to the church. The lady in the house told me that the door was already open for the day, so I wandered back. Sure enough, the door was unlocked. The little building was very simple, with a few small plaques on the wall and a book in which to write one's condolences. I thumbed through the book and was amazed to learn that people from all over the world had visited the site.

I stood there for a few moments, thinking of all that one might enter into this book. It was November 5, 1996, and President Clinton was being inaugurated in Washington, DC for his second term. Franck Curtin had told me that President Clinton knew the truth about the Boeing 747's. Clinton was keeping quiet, and Curtin was keeping quiet. What should I write? I decided to remain dignified, and I simply signed the book.

I had all of the information I needed, so I decided to head to London to buy some of the books I had identified in the Lockerbie library. You would have thought that one could buy any book you wanted in London. In the area known as Charing Cross, the famous Foyles bookstore is located along with dozens of others. But I could not find any of the books and none of the people working in the stores was interested in helping me find them. Did I say that I liked Brits?

I left for Geneva to pick up the suitcase I had left there, full of irreplaceable documents, and to devote a little time to my business. (Originally, I had gone to Geneva to evaluate opportunities for an American company to enter the Internet business in

Switzerland, thinking — incorrectly — that it might be substantially behind the USA.) I was nervous. I tried the key on my locker at the station, and it did not work.

This could turn interesting. What if I were to ask the porter for my case, only to find myself confronted by someone who opposed my sleuthing efforts? I worried unnecessarily, as the porter quickly opened the locker (after I had described what he would find, and five francs, to boot).

I rented an office at Place de Cornavin, opposite the railway station, and quickly took my cases to the room. I locked them into a closet. I checked into the Montana Hotel, around the corner, and settled down to work. The Montana is not a luxury accommodation, but I was very comfortable in Room 67, where I would spend many days. I became friends with Mr. Mueller, the owner, who helped me to master the details of working in Geneva. Throughout my stay, I kept notes and drafted sections of this book in my spare time. Meanwhile, I had been keeping in touch with Franck Curtin, to make sure that he sent the Boeing Report to Dardas.

The next thing I knew, I had a fax from Curtin demanding another $10,000! I was really upset by this apparent blackmail, but I assured him that everything would work out, and I left it at that. I could not risk annoying him and not getting the report as agreed. I decided I would not tell Dardas about the latest demand, but to wait and see what would happen. That turned out to be a good decision, as Curtin sent Dardas a copy of the Boeing Report without question. Dardas is a lawyer and would take a very dim view of being screwed by a Frenchman.

Meanwhile, the papers were full of the TWA 800 investigation and each day I scoured the news for information. I finished my work in Geneva and went back to London with my suitcases filled with information about Boeing 747 crashes. I found a very large leather suitcase, with a strong lock, and loaded all my data, videos, tapes, models and books into the case. I tucked it away for safekeeping in a friend's apartment, off the Bayswater road. And I headed back to Houston to continue my search for the truth.

Chapter 5

EVERETT, WASHINGTON. BIRTHPLACE OF THE BOEING 747

My investigation was taking on a life of its own. The pictures of the memorials at the Dryfesdale cemetery and Tundergathen were vivid in my mind, and images of the plane falling from the sky began to haunt me. I had a moral obligation to do the right thing with all this information. But what was the right thing? To expose what I knew, as soon as possible, or to gather exhaustive and conclusive information?

I was driven to find the truth; yet, I felt that I had to avoid alerting anyone to my activities, although that would, to a degree, limit my options. I went through the entire library system in Houston and was amazed at how much I could find. There were movies, documentaries, books and articles on the Lockerbie crash, but none of it was organized into any methodical arrangement. The information had arrived at the libraries, but no one was taking enough interest to cross-reference the material.

Even before the crash of TWA 800 that July, the TV news

(particularly CNN) had a constant stream of relevant information, which I carefully recorded on my VCR. Of particular interest was a program aired on the Arab Summit meeting in Cairo. Colonel Gadhaffi had violated the sanctions and flown to Cairo for the meeting. After the meeting, Christiane Amanpour interviewed King Hussain, Hosni Mubarak and Colonel Gadhaffi. I recorded all the interviews.

Gadhaffi was very amicable, with his usual soft-spoken voice (when not exciting his followers). He was clearly sending a message to President Clinton, whom he said he respected. He said he bore no grudge against the American people, who had been subjected to a very heavy propaganda barrage over the years. I was particularly interested in the fact that he appeared to be sending a message to President Clinton.

By sheer coincidence, my wife and I had been making plans for a business trip to Vancouver, to look into the venture capital markets for business development projects. It is very expensive to fly almost anywhere in Canada; but it is relatively cheap to fly to US towns near the border. Also, we were approaching retirement age, and were beginning to consider where would be the most ideal place to move to. To me, retirement meant working on projects, as I always have done, but at a more relaxed pace. To do that, I would need to be in a place that was attractive but, at the same time, not too far from a major transportation and information center. We have traveled all over the world and many places came to mind, as we pondered the question; but we were also learning that the choice of location could be very tricky. We had watched our parents get it wrong, locating themselves in places that made it difficult to get to shopping areas, with no interesting and easy places for walking; and worst of all, there was no place to go to meet new people and keep life interesting. In that case, the next thing you know, you are watching the circus on TV!

We were not about to make the same mistakes, and we were going to begin the evaluation well in advance. We were both brought up in England, so we had a natural affinity for the sea. We loved Canada with a passion, especially the Rocky Mountains, and we were not built for living and working in a hot, humid climate like Florida. The Pacific Northwest has always appeared to be very attractive to our retirement aims and so we decided to take a busman's holiday/business trip to Washington State. We had at least six other places in mind to evaluate in the summer of 1996, and it was pure coincidence that we chose to go to Washington State first. I had not intended to visit the Boeing plant at Everett, but after my trip to Lockerbie, it was certainly on the itinerary.

We arrived in Seattle early in August 1996 — shortly after the TWA 800 crash on July 17. The crash was big news in Seattle and in Everett, home of the Boeing 747, where thousands of people work at the plant and its subsidiaries. Needless to say, the local newspapers were a major source of information on the TWA 800 crash.

We spent a couple of days around Seattle, and decided to head north to Canada, visiting the Boeing plant *en route*, on Monday, August 26, 1996. I just assumed that plant tours would be available. I doubted there would be many visitors on a Monday.

I could not have been more wrong. The tours were almost a separate organization, run by retirees who could not stay away. If you love planes, as I do, you will understand the feelings those people have about these remarkable machines. We arrived with the expectation of being able to start touring immediately; to our amazement, we were told that all the tours were full for the day. People from all over the world had booked tours, such was the fascination of this giant operation and it included all types of people.

Entrepreneurs are optimists by nature, and we never take "no" for an answer. I asked Jan if we could just hang around, and see if someone would tire of the wait and go home. I told the old fellow who was in charge of loading the buses that I would wait for hours, if I had to: I just had to see the plant. He was very amused by my approach and I took every opportunity to ask him about his life with Boeing. He was falling for the bait. Next thing I knew, there was a special bus arranged for the diehards. I was fascinated as I remember visiting the de Havilland plant in Hatfield, where my Dad had spent some forty years making and inspecting some of the greatest planes in history. Now I was involved with a similar story some fifty years later.

The tour was very controlled, as you can imagine. We were guided by one person and watched by several more. There were no cameras, which was unfortunate, but I knew what to look for and felt sure I would be back some day. I was quite sure the whole trip was being monitored and recorded in case, for instance, someone decided to fake a fall and set up a lawsuit. We were taken down one corridor in the center of one of the largest buildings I had ever seen. I believe the corridor was one mile long. We walked about ten minutes towards the center of the building, and then the entire group was put in an elevator to a viewing area. As I stepped off the elevator, I saw a cross-section of the Boeing 747, showing the size of the outer fuselage and a series of seats attached to a cross member. I touched the outer shell and even I was amazed at how little metal constitutes the "skin" of the Boeing 747. I have seen many planes in construction and I know that the designers have to meet certain specifications with a considerable margin of safety. One the other hand, when one has a catastrophic failure such as the break up of the TWA 800, one can see why it ends up in so many small pieces.

I turned to the left and walked along the gantry walkway from which I could observe the manufacturing of the Section 41's. There were about four in a row with workmen busily assembling this complex piece of machinery. It reminded me of the method of building the Comet. Of course, the technology has advanced incredibly and the actual details of the systems, the fabrication techniques and the materials are quite unlike their predecessors. I remember walking around the Comet plant at Hatfield and being amazed at how dirty and unorganized the machine shop had appeared. The Boeing plant was spotless, with notices on the walls to keep reminding the workmen not to lose tools or parts on the plane, which might cause problems down the road. It appeared to be an extremely well-organized operation.

I spent some time studying the Section 41, as I hoped one day to be writing this narrative. To the left, there was a huge assembly bay where the fuselage was being married to the wings and Section 41 and Section 42. I would have liked to stay longer and wandered on the assembly floor, but the tour was tightly organized and we were limited as to the time we could watch the plane being assembled. I asked the guide if Boeing did any remanufacturing or service work and his response was negative. The company only provided parts and technical support. Each airline was responsible for its own repairs. Finally, I learned that Boeing has about 120,000 employees directly on its payroll and thousands of subcontractors all over the world.

After the Boeing plant tour, we left for Vancouver, visiting some of the nicest towns along the way. We had seen Edmonds, earlier, and liked it very much. We visited Mount Rainier, and stayed in a log cabin on the way down. It was so beautiful and relaxing and yet I was constantly aware of the hundreds of people who had had their lives cut short, thus denying this experience. Even though this was a "pleasure trip", the process of collecting information was never away from my mind.

In Vancouver, we visited friends from our Libyan days; I was dying to relate my story but I knew that "loose lips sink ships". We took the ferry to Vancouver Island and took long walks in the forests to seek out rivers and wildlife. The salmon were heading up stream and we spent many hours watching the fascinating process. Nature works in strange ways and one has to wonder what drives man to compete in his world to the detriment of others. Like the salmon, people live their lives knowing they need each other — but quite willing to accept the demise of others.

Victoria in the summer is a delightful place to stay. We had

Everett, Washington. Birthplace of the Boeing 747

Above: The first flight of Boeing 747. The flight went smoothly until a problem with a misaligned segment of an inboard flap section forced an early return to Everett. Photo: Boeing)

Below: Everett, Washington. Section 41 about to be joined to the main fuselage.

been living very modestly until we arrived in Victoria and decided to have tea at the Empress Hotel. This is a snobby little tea, based on the old British tradition in the colonies. Having been raised in England, Jan and I found it amusing to see the old folks hanging onto their past via the tea, sandwiches and little cakes, very pleasing to the palate. Each day I bought all the newspapers and carefully reviewed them for any new hints.

Chapter 6

GATHERING EVIDENCE: TIMING DEVICES

My intention in writing this book is to offer certain new facts and insights that have resulted from my investigations; it is not my goal to reiterate the voluminous data that is already published. Still, some background information must be included in order to maintain a logical, coherent flow and to enable readers to see how the new information fits in with the rest.

There was no question that I was on a crusade. The time that I was spending on this fact-finding mission was hurting my business. I checked the Internet every day, reviewed the information and made copies of any relevant facts. The stories were amazing. Clearly, there were many other amateur investigators working on the TWA 800 crash, and a few people were looking at the relationship between the various crashes. One person in particular was getting close to my evaluation, even without the ace I had up my sleeve, the Boeing Report. I was tempted almost every day to begin posting my work to the Internet discussions, but I still thought that I was in a unique position and should make sure I had all my facts straight, before letting loose with conjectures.

Surprisingly, I have not seen one article by a Boeing 747 pilot. They are surely all talking about the situation, among themselves. Can they have been ordered to stay out of the public discussions, for fear of igniting public concern over flying in these planes?

As the months dragged by, spokesmen from the FBI, the FAA and the NSTB (National Safety Transportation Board) were constantly on the TV news, voicing their varied versions of what might have happened to TWA 800 and what they were doing about it. Government agencies, the Navy and dozens of private boats were out on the ocean every day, gathering pieces of the wreckage in what has turned out to be the most expensive investigation ever conducted on an airline crash (the total cost will probably exceed $150 million).

If my information is correct, then this effort was just a public relations stunt. Certainly, the French are very upset by the apparent deliberate side-stepping on the facts; their dismay can be seen in the French newspapers and magazines. A large percentage of the TWA 800 victims were French, and their families had the same right as the Americans had, to know what was going on.

By November 1996, the investigators were considering three theories: terrorism, a missile (friendly or otherwise), and mechanical failure. The authorities and dozens of amateurs were pursuing each case with the utmost fervor, and much of their findings and feelings were being expressed on the Internet. (There are so many files on the Internet that it has become almost impossible to glean fact from fiction, gossip from enlightenment, and "lack of naivety" from "paranoia".)

I watched hours, days, of news and documentary programs. In several instances that I can identify, the basis of the argument offered was clearly out of line with reality. After TWA 800, another crash occurred off Peru with a Boeing 757; these are different types of planes, and a 757 has nothing to do with the Boeing 747 of early vintage.

The Boeing 757 investigator's report was total nonsense and was very disturbing. It suggested that the pilot had left a cover on the pitot tube, the key component of the instrument measuring air speed. How does the anemometer (air speed indicator) work? You may have seen the tubes that protrude in the front of a fighter jet. These are pitot tubes. They are placed on the aircraft in a location where the airflow is not be disturbed or distorted by the passage of the plane through the atmosphere; the front of the tube is directly exposed to the oncoming air. The pitot tube has a hole in the front, through which the oncoming air exerts pressure inside the tube. (Just as, if you hold your hand out the car window while underway, you can feel the pressure pushing it backwards.) On the side of the aircraft, in another undisturbed area, is a tube called the static line. This measures the normal pressure at whatever altitude you may be flying. The two are read, together, by the an instrument located in the cockpit.

The calculation of air speed on the basis of the difference

between the pitot tube pressure and the static pressure is very accurate, and such instruments have been in use for over fifty years. No one would fly without an airspeed indicator. These instruments are critical for safety, and are commonly protected by a hood on a plane that is parked, so that nothing can get into it during "down time". This hood has a long red band that will stream from it, if it is left in place inadvertently; the pilot will always check to see that it has been removed before taxing. The ground personnel, too, would easily notice its presence when guiding the plane away from the ramp. When a pilot is on the taxiway, he always checks the instruments as the plane moves forward; many instruments respond to even small changes of speed and direction. When the pilot is taking off, he is basically only interested in two things: looking out the window and looking at the airspeed indicator. To oversimplify, when the plane is going at the correct speed, the pilot pulls back on the yoke and lets the plane take off.

Now, for the investigator of the Peruvian crash to suggest that the pilot had left the hood on the pitot tube is what my former mates would call rubbish. The pilot would have seen the problem seconds after the initial roll down the runway and he would have aborted the take off. There are many such reports by investigators who manage to get an audience on TV; they spew out technical garbage, which they clearly do not understand, and it does not serve the flying public well.

Any number of qualified people have conducted painstaking evaluations of airplane crashes, and many have written books with detailed descriptions of their understanding of the facts.

Gathering Evidence: Timing Devices

Several of them provide evidence supporting the "Section 41" theories about the earlier Boeing 747's (see the bibliography). Indeed, I plan to work toward opening a public reference center to facilitate research using these materials, which can be hard to find.

I was tempted to contact many people who I thought might be in a position to add to the story. Then, each time, I decided that most people could not be trusted to keep the story quiet, to resist the temptation to "beat me to the market", or even have to my work subpoenaed by attorneys with their own agenda. My business life has taught me to keep information tight until the time is right.

Anyway, Dardas had a contract, which could be very lucrative, and there was no way we were going to jeopardize that opportunity. In the north of England, there is a saying (which sounds even better in the Yorkshire pronunciation than in modern English): "Hear all, see all, say nothing; eat all, drink all; pay nothing. If you do anything for nothing, do it for yourself." This was a case in point.

By the end of 1998, I was watching the timing of the upcoming trial of the two accused Libyans. Initially, the suspects were in jail in Zeist, Holland; according to Scottish law the trial had to being within 90 days. Then the defense lawyers obtained a special ruling on June 7, 1999, allowing them six more months for preparation. The trial was to start February 4, 2000. Then the trial was delayed a second time, to May 3, 2000, and I had to wonder why. The prosecution wanted to avoid any opportunity for the defense to cry foul! After all, the prosecution had had 10 years to prepare, and Libya had introduced a new lawyer, Kamal Maghur, who needed time to become familiar with the case.

This gave me time to do a final run at the research and to take a few more risks. The risks may have been imaginary, since everything I had done was legal; still, I was concerned that I could be stopped in my work by some kind of legal challenge.

There was one concern in particular. When it come to anything negative about the Clinton Administration, there is a spin team with unlimited funds ready to go on the offensive — and there are no rules of engagement. According to sources on the Internet, over 50 people who, at some point in their lives, had run afoul of Clinton were now six feet under. The most interesting one was Vince Foster, supposedly a one-time lover of Hillary Clinton. Numerous books have been written about Clinton in Arkansas; they make for chilling reading. In *The Secret Life of Bill Clinton*, by Ambrose-Evans Pritchard of the UK *Daily Telegraph*, I found a reference to the fact that, although he had completed most of his research and interviews in the US (mainly in Arkansas), he decided to write the book in an isolated area of Kent. I took his advice, and chose to write much of the current work in the Hotel Alpes et Lac, Neuchatel and in the Hostellerie des Chevaliers in Gruyere, Switzerland; and I finally finished the first draft in Flat 5 of the Suncourt Holiday Flats in Cromer, Norfolk. Now, I decided to take a few risks.

Timing Devices, or Doubtful Evidence

Edwin Bollier lives in Zurich, Switzerland and ran a company called MEBO Ltd. Mr. Bollier is (was) a key player in the

Pan Am 103 bombing.

Mr. Bollier ran a small electronic communications design and manufacturing company, from his shop at 414 Badnerstrasse, Zurich. When the FBI found a very small fragment of the timer that allegedly set off the bomb on Pan Am 103, they "found" the name MEBO inscribed on the circuit board. I raise a question mark here, because the lettering was almost unintelligible and the FBI investigator supposedly deduced the name, MEBO, by checking all other possible manufacturers of printed circuit boards; this is an extremely weak way to approach the analysis. My next book should focus on this story, alone.

This fragment, claimed the FBI, was the key evidence of a bomb having brought down the plane over Lockerbie. Furthermore, it was used to link Libya to the dastardly deed.

Bollier was immediately sued for $32 million by the Pan Am Insurance Company as a participant in the bombing. Bollier's bank credits lines were closed and his business collapsed.

MEBO offered, over the Internet, to pay $10 million for any new information on the Pan Am 103 bombing. Through my own "back channels" we contacted Bollier on November 25, 1998 and arranged a meeting in Zurich. Bollier offered to pay the expenses but I declined; I wanted to avoid accepting any potentially compromising ties to anyone.

Bollier's offer of $10 million was a sure-fire magnet; he received emails from people asking if they could simply help, from university students writing research papers, from cranks and con men — and from a retired Boeing employee.

The Boeing employee was broke and wanted cash upfront,

which was a clear turn-off to Bollier. The fellow did hire an attorney, Mr. William McKinnon, to represent him. Curiously enough, this Boeing insider had internal memos and information, which I thought could be useful, so I took a copy of his e-mails with the intention of contacting him later. And later in my tale, you will see what came of that.

On December 15, 1998 I met with Ed Bollier at the Intercontinental Hotel, which stands next to his factory. Bollier is a German Swiss. I do not speak any German, but fortunately Bollier's English, while not perfect, was more than adequate to exchange ideas. He was clearly interested in what I had to say and we quickly developed a congenial rapport. Technical people usually find common ground and Bollier and I had little difficulty reviewing the technical aspects of the case against him. I was fascinated. Bollier had lost his house and business, and to clear his name he had spent eight years on the Pan Am 103 investigation. I doubt there was one single aspect of the Pan Am 103 disaster that Bollier and his attorney did not know; more than you can ever imagine.

I saw a massive amount of the evidence — but not all; Bollier would not let me see it all — just enough to trade fact for fact between us. Bollier told me I had seen about 40% of the information and could have the balance once the trial was underway. I intend to take him up. There is so much to this real-life story that I do not understand why people read fiction.

Our first meeting at the end of 1998 was really a "get to know you" session; it was followed by three more meetings and weekly calls. The more we discussed the situation, the more comfortable Bollier became; he was quite candid when he said that at first, he'd thought I was from the CIA or even Mossad! I assured him I was neither that clever nor did I have unlimited funds. We became friends.

The Timer: Bollier was dragged into this due to the fact that his company, MEBO, designed and manufactured small quantities of an electronic timer that could be employed in the detonation of explosives. Three years before the Pan Am 103 crash, Bollier was working with the Libyan military and was selling MST-13 timers. Bolliers' technician hand-built three prototype timers. The first one had problems, but was later modified and fixed; numbers two and three were OK. The timers were made from an initial film of the circuit.

In these early days Bollier, with the consent of the Swiss authorities, visited East Germany (before the Berlin wall came down) to demonstrate his timers. He took the first three examples to the Institute of Technical Research, which was related to the East German Secret Police, the equivalent of the FBI (the Stasi). The East Germans finally bought seven timers: the first two prototypes and five new ones. Then a curious thing happened. On July 16, 1988, six months before the Pan Am 103 crash, the MEBO offices were burgled and prototype Number Three was stolen, along with two films. Fortunately for Bollier, he reported the theft to the Swiss police. It is on record.

Bollier was then working with the Libyan military, and eventually sold them 20 identical, carefully manufactured, timers. All of this activity, Bollier learned later, was being observed by the CIA in both Libya and Zurich.

One Charles Byers has asserted that the MST-13 timers were, in fact, manufactured for the CIA in Florida. In a letter to the Honorable Portor Goss, Chairman, House Intelligence Oversight Committee on February 14, 1998, Mr. Byers states that he specializes in ordinance, explosives and pyrotechnics though his company, Accuracy Systems Ordinance Corporation. In his letter, he suggests that the Arrow Air DC-8 that crashed in Gander, Newfoundland on December 12, 1985, (www.dcia.com/ganderl.html) had run afoul not, as the official version claimed, of icing on the wings, but had been bombed out of the air, killing all 248 servicemen and eight crew members. There is a picture of the blast hole in the aircraft to prove his point. He points out that his firm designed, manufactured and sold the bombing device exclusively for the CIA.

The "proof" that the bomb-timer was a product made by MEBO was found by a Mr. Thomas Thurman, (not a technical but only a political scientist) at the FBI laboratory. Thurman had no professional training in explosives investigations, yet he had been assigned to the FBI laboratory (which has been under considerable criticism for sloppy and even criminally conducted analysis). Criminal, because it was falsifying tests in a deliberate attempt to obtain convictions that it could not otherwise prove. The FBI's sloppy work was brought to the attention of the general public in the cases of the Oklahoma bombing and the O. J. Simpson cases, among others. Thurman's analysis of the MEBO timer was even more egregious.

The whole case against the Libyans rested on his work; but he had never even seen the fragment of the timer, only a photograph.

Even Bollier, who was being accused of indirectly aiding the bombing, was not allowed by the authorities to see the fragment, despite many attempts by his attorneys. Why was that? When Bollier saw the picture, he knew right away that there were serious questions about the timer.

The timer in the photograph was, indeed, similar to those produced by MEBO — with some significant differences. The fragment, no larger than a thumbnail, was of the touch pad on the timer. In an interview on the British Channel 4 TV program, Bollier placed a photograph of one of his timers alongside the Thurman timer; major differences were apparent. One part of the circuit did not go to the edge of the board, on the Bollier timer, but was clearly at the edge on the Thurman timer.

Thurman accused the Scottish police of having cut off the top of the timer that he had been examining. This was a very serious accusation; if true, it would have amounted to tampering with evidence. The case could have been dismissed. For the Scottish police to have cut this small fragment so cleanly would have required a very deliberate effort, with the intent to alter the evidence, and it would have required a very special cutting tool to make such a clean cut.

On one side of the fragment, there is a curved area that has been carved out to allow the timer to be fitted into the timer case. On the Bollier timer, the curve is smooth and regular, as one might expect in a professionally-made circuit board. On the Thurman timer, one can clearly see the tooth marks of a saw. Bollier told me there are other features about the Thurman timer that he was not prepared to discuss, pending the trial. I believe it has to do with the soldering on the board; it is possible that the Thurman fragment was never even made into a timer.

The authorities have officially criticized Thurman for having fabricated evidence, and for sloppy work. He now teaches explosives at East Kentucky University. He is a crook, and will be called as a witness in the Libyans' trial. We will pickup this strand in the story in the final chapter.

The Two Accused Libyans: Abdel Basset Ali al-Megrahi had known Bollier for some time. Megrahi had leased an office from Bollier on the same floor of the Badnerstrasse office building where Bollier had his office and factory. Bollier was not at all pleased when his Libyan tenant quit the office without paying the last bill. He was apparently working for the Libyan government and, since Switzerland was the only European country with a Libyan embassy, he might indeed have been conducting normal business.

Leman Khalifa Fhimah worked for Libyan Arab Airlines in Malta and, according to documentaries that I have on video, he was not opposed to doing a little business on the side. One such arrangement that he sought was to obtain a contract with Malta Air to print their luggage tags. Fhimah had access to the entire airport, including the baggage loading system.

On November 14, 1991 the prosecution authorities in the US and Scotland simultaneously brought criminal charges against the two Libyan, accusing them of bombing Pan Am 103 and working for the Libyan intelligence service. This was a complete surprise to everyone who had been following the case.

Vincent Cannistraro, who was in charge of the Lockerbie investigation for the CIA, told Morley Safer (of 60 Minutes) in early 1991 that, "We believe — and I think it has been accepted by the President's (Bush) Commission on Aviation Security — that Ahmid Jibril, a leader of the Palestinian Liberation Front, ultimately orchestrated the operation that led to the destruction of Pan Am 103 in December 1988."

When Iraq attacked Kuwait, Bush decided to orchestrate a war against Iraq; it became known as the "Gulf War." President Bush and Prime Minister Margaret Thatcher (UK) needed the support of a coalition of countries for this war, including Syria, which had been initially accused of working with Iran on the bombing of Pan Am 103. Coincidentally, it had not gone unnoticed that Bush — who now needed the support of Arab countries who, ordinarily, were less than friendly with the US.

According to the New York Times and other newspapers, President Bush had met with President Hafez Assad to discuss Syria's possible contribution to the multinational task force confronting Saddam Hussein in the Gulf. This scenario prompted President Bush to remark, "The Syrians took a bum rap on this." It then appears that Bush decided to blame Libya for the Pan Am 103 bombing.

Cannistraro told the *New York Times* that it was "outrageous" to pin the whole thing on Gadhafi. Cannistraro subsequently changed the whole story, and now he, too, blames the Libyans. Bollier also gave me a copy of a report by the Israeli Mossad, wherein President Bush is accused of ordering the assassination of a Mr. Amiram Nir for having secretly taped Bush's conversation with President Assad of Syria.

The FBI then placed the two Libyans on their Ten Most Wanted list; but Gadhafi refused to give up "one of the sons of Libya". Things came to a standstill until early 1999.

The Toshiba Radio: This is another endless strand in the Pan Am 103 story: the harder you pull, the more longer and knottier it appears. Its tentacles spread over the entire problem of the liberation of Palestine; the question of war and terrorism will come up later in the book.

The Americans, British and the French were instrumental in creating the state of Israel in 1948. The Jews had suffered, as we are all constantly reminded, and it made sense to create a state where they could live and prosper without further persecution.

Let it be said that there had been many wars between Israel and the Palestinians since then, with many casualties. The Palestinians could not confront their enemy head-on, especially with the awesome might of the US military behind the Israelis. The only solution was a form of guerrilla warfare, which has now become known as terrorism.

In the late 1990's, the German secret service, the BKA, had been carefully watching a group of Arabs operating in West Germany under suspicious circumstances. In the first two years of the investigation, it was widely accepted that the bombing of Pan Am 103 began on July 3, 1998, in the Persian Gulf. While sailing in Iranian territorial waters, the US Aegis-class cruiser Vincennes "mistook" an Iranian Airbus for an Iranian F14 fighter. The Iranian Airbus was shot down by a missile, with the loss of 290 innocent people, most of them on their way to Mecca.

The Iranians were incensed when the US Navy commended the captain and the crew for protecting the battleship. The Ayatollah Khomeini ordered the destruction of four American airliners, but it was to be carried out discreetly.

On July 9, 1988 Ali Akbar Mohtashemi, who was in charge of securing revenge, awarded a contract to Ahmed Jibril, a former Syrian army officer and head of the PFLP-GC. He paid him $10 million; the CIA has traced the money to accounts in Switzerland and Spain.

Jibril chose the Frankfurt airport because Germany had a very substantial Muslim community, including Islamic fundamentalists, and some of the baggage handlers at the airport were Turkish. Frankfurt was an important hub for American carriers and served as a connection point for traffic throughout all parts of Europe and the Middle East. Jibril developed a team of likeminded Arabs including Hafez Kassem Dalkamoni and Abdel Farrah Ghadanfar, and set them up in Neuss, Germany. On October 13, 1988, they were joined by Marwan Abdel Khreesat, a Jordanian TV repairman and the leading explosives expert and bomb-maker for the PFLP. It is understood that Khreesat was an informant to the CIA and, indeed, the Israeli Mossad. The German police were keeping watch over all of the participants, recording all of their phone conversations and amassing a significant file of photos, videos and other documents from all over Europe and the Middle East. The code name for the operation was "Autumn Leaves."

Dalkamoni and Khreesat were filmed purchasing electronic equipment at local stores. From these components, Khreesat was able to make four or five bombs in Toshiba Bombeat 453 radio/cassette recorders. On October 26, 1988, the police began a 24-hour raid on apartments and houses in five cities and arrested 16 suspected terrorists, including Dalkamoni and Khreesat. Interestingly enough, Dalkamoni and Ghadanfar were held on terrorist charges and the others, including Kreesat, were released "for lack of evidence". They disappeared.

The Toshiba Bombeat 453 radio, found in Dalkamoni's Ford Taurus at the time of his arrest, had 312 grams of Sentex-H explosive molded into the case with a barometric and timer delay switch. The same type of radio was supposedly found in the debris of Pan Am 103, which linked the bombing to Jibril.

A person who is sophisticated enough to build such a bomb and get it planted on an airplane at an international airport must be assumed not to be a fool. We must study the alleged course of action and see whether it makes sense.

If one intends to bomb a US jumbo jet, it would be best to arrange for the blast to occur well out over the Atlantic Ocean, where the recovery of the debris would be very difficult; one would certainly not intend to cause the detonation while the plane was still over land. Thus, if indeed the Libyans had placed the bomb on the plane while it was at Malta, the perpetrators would have had to design a bomb that would not explode on the way to Frankfurt. It would have to survive the trip to London Heathrow, and then explode some place over the Atlantic. This bomb would also have had to pass undetected through several airport scanners. Planes from Malta to Frankfurt and Frankfurt to London fly at at least 30,000 feet during some part of the journey, which would have activated the barometric switch. Clearly it was not.

Maybe the electric circuit was designed in such a way that the timer required to be activated before the barometric switch. This would have been very complex electronics, unique for the situation, and a very curious bombing program when a much simpler design would have accomplished the same goals. This complex bombing program would be made even more difficult by virtue of the fact that many planes leaving Heathrow in the winter weather, and especially just before Christmas, could easily have been delayed for hours. I know: I have sat on the runway for up to three hours myself, waiting for clearance, and only a few of those cases involved technical problems.

Let us assume that the Libyans had been given explicit instructions; certainly, those who implement these scenarios are not usually the same people who design the game plan. The planners must have known that Malta Air Flight KM 180 would leave the Valletta airport at 10:15 AM and that Pan Am 103 was due to leave Heathrow at 6:00 PM the same day, December 21, 1988. That was a time span of 7 hours 45 minutes. The trip across the British Isles would be approximately 45 minutes. Hence, the plane could be expected to be near land for 8 hours and 30 minutes. The trip across the Atlantic to New York was probably five hours; so the best time to detonate a bomb in the plane would have been between 7:00 PM and midnight GMT, while the plane was over the water. Allowing for a potential one-hour delay, a wise bomber would set the timer to go off between 9:00 PM and 11 PM. The best time would have been 10:15 PM GMT, exactly 12 hours after leaving Valletta. To do the job right, the window of opportunity was at least two hours — a fairly simple analysis for the bomber. Why, then, did the explosion occur at 19:03 GMT, only eight hours into the long journey? That is well outside the margin of error.

It would have been far more sensible to load the bomb at Heathrow and the Iranian baggage handling facility was immediately next to the facility loading Pan Am 103. The plane was 25 minutes late. Had it left on time it might well have been over the ocean! Even so, the flight over the ocean takes about five hours and it would have been more effective to time the bomb to go off in about three hours instead of about one hour into the flight from Heathrow.

I have never seen any evidence indicating that the Toshiba radio had a barometric switch like the ones built by Kreesat in Germany; however detailed, my documentation cannot be considered totally complete. There is a major controversy over whether or not the bomb on Pan Am 103 used a combined barometric and timer switch, like the Kreesat models, or just a timer. The Libyan/Malta theory rests on the supposition that only a MEBO timer was used. There are many holes in the evidence as presented to the press.

The Toshiba Bombeat radio-cassette was identified on the basis of a very small fragment of the circuit board; that's another amazing find, considering the debris was scattered over 845 square miles with the full force of a Semtex-H explosion — which I would have thought would vaporize a circuit board. If, indeed, a small piece survived the blast, then that is the fragment, along with the timer fragment, that served as the basis for switching the charges against Jibril to the Libyans.

Meanwhile, Bollier was not sitting on his hands. He had a $32 million lawsuit hanging over him; he'd already lost his business and indeed his house. Bollier is an electronics expert and after an in-depth analysis, he found that the radio-cassette player fragment was from a Toshiba, alright — but from an RT-8016 model, sold only in the USA! The identifying part is symbol number L106, part number 22245500, description Coil, MW oscillator. Now that is going to take some explaining.

The Suitcase. The suitcase that is supposed to have held the Pan Am 103 bomb has a story of its own. In my review of the available literature, a brown Samsonite suitcase is mentioned. It is said to have been brought into Germany from Jordan, by Kreesat. Unfortunately, when the German police's "Autumn Leaves" operation went into effect and they raided the Neuss apartment, and the Ford Taurus car, as Kreesat was leaving (having completed his work), there was no sign of the suitcase. It seems strange that Dolkamoni and Kreesat, who clearly had no prior knowledge of the raid, would have hidden his suitcase; this is circumstantial evidence at best.

Furthermore, how would this suitcase have gotten into the hands of the Libyans in Malta? And why would Khreesat send his own suitcase to Malta, when it would have been easy to purchase one locally? That is still a mystery!

How did the police know Kreesat had a brown Samsonite suitcase, if they could not find it? Was a photograph taken of Kreesat as he left the Frankfurt airport to meet Dalkamoni?

Drugs. Now we turn to the smuggling of drugs from the Bakaa valley, Lebanon.

At least two other brown Samonsite suitcases were supposed to be on flight 103. Apparently, a Pan Am pilot based in Berlin had done his Christmas shopping and loaded two identical brown Samsonite suitcases with gifts, in preparation for the trip back to the Seattle. At the last minute, he had to take a flight to Karachi. He therefore arranged for his two suitcases to be flown straight from Berlin to Seattle on Pan Am 103. The suitcases went through Frankfurt — but only one was found in the wreckage of flight 103, and the second one arrived the next day in Seattle. This story was exposed in Time in 1992. This underscores the fact that the Pan Am baggage system was far from satisfactory; and it did not conform to FAA regulations at the time: unaccompanied bags were clearly being transported, with only automated scanning to ensure that no bombs were included. And secondly, it makes it clear that brown Samsonite suitcases were very common.

The Royal Armament Research and Development Establishment (RARDE) operating at Fort Halstead, in Kent, had been tasked with doing forensic work on the debris from Pan Am 103. On a piece of the luggage container, it found traces of SEMTEX-H explosive material, and tests on the luggage itself showed that the bomb had been placed in a copper-colored Samsonite suitcase.

Bollier, in his detailed investigation, found a startling fact. The suitcase that was identified as the one carrying the bomb was not a European model, like the one assumed to have been purchased by the Jibril group. It was a US model, made in Denver, Colorado!

The Suitcase, Bakaa Drugs, Iran-Contra Affair. This next section is based on information gathered by others and published in documentaries and books that have been banned or censored by both the British and American government authorities. There seems to be a belief that the British and American publics are so incapable of handling complex topics that they have to be protected by government employees, who are better suited to judge what is true and what is bogus.

Of all the documents that I scoured for the following discussion, the most concise work is a book by Donald Goddard with Lester K. Coleman, entitled *Trail of the Octopus: From Beirut to Lockerbie-Inside the DIA*. In addition, the documentary called, *The Maltese Double Cross* is highly regarded. Write to your political representatives and tell them you want access to both documents!

I watched a good deal of the Iran-Contra US Senate Hearings, which involved a very complex story of how President Ronald Reagan and Vice President George Bush funded the war in Nicaragua after congress had refused to fund the effort. I would have been a great supporter, had I been asked, of Lt-Colonel Oliver North, if only he had not spent days and nights shredding documents before the congressional hearings.

What was he hiding? North was "working" for Robert McFarland, Reagan's National Security Advisor. (McFarland later had a nervous breakdown over the affair, and tried to commit suicide.) It has been suggested that Pat Robinson, who headed up the Christian Broadcasting Network, was also heavily engaged in raising money in Lebanon to support the Nicaraguan Contras. He was working through Major-General John K. Singlaub, president of the World Anti-Communist League and through Ollie North, with the covert assistance of William Casey of the CIA. (In *The Marcos Dynasty* by Sterling Seagrave, you will find Singlaub and a few other well-know operatives trying to sell billions of dollars worth of gold that Marcos had stolen. That was used to fund the Contras, too.)

There were two major covert political operations going on at the same time: one to fund the Contras, and the other to have hostages released from Iran, in return for arms and/or drugs. In the 1986's, Oliver North and Robert McFarland were secretly organizing the sale of TOW missiles and launchers to the Iranians in exchange for the release of the American hostages. When Beirut's Arabic news magazine broke this story on November 3, 1986, it caused an international scandal of such magnitude that President Reagan fired North and Rear-Admiral John Poindexter, who had replaced McFarland as his National Security Advisor. Reagan became mired in Irangate. This did not stop North being further involved with the release of the hostages and other unsavory business in the Middle East with the CIA.

The Interfor Report, commissioned by Pan Am, was based on investigations on the work of Mr. Juval Aviv, who was said to have been a member of the hit teams that killed every member of the Black September Group (responsible for the Munich massacre of Israeli athletes at the Olympic Village). He was extremely well connected to Mossad and other intelligence agencies. According to the Aviv report, in the summer of 1988, Matthew Kevin Gannon (the deputy chief of the CIA) and Major Charles Dennis McKee went to Beirut to prepare for the possible hostage rescue effort. The French had recent obtained the release of their hostages from Iran by selling arms to the Iranians. McKee, a member of the US Defense Intelligence Agency (DIA), had a brown Samsonite suitcase carrying $500,000 cash, with maps and photographs of where the hostages were located. They were prepared to get the American hostages by whatever method it took, including force.

Then I found a startling inference that the suitcases carried by McKee were probably wired with a bomb big enough to destroy any information in the event it was opened by anyone but him. Could this have been the case? Did it go off accidentally? Why didn't the Scottish police find more parts of the bomb? After all, the Toshiba radio would not have completely vaporized.

Rifat Assad, brother of President Hafez Assad of Syria, was the undisputed overlord of the Lebanese heroin purification laboratories, which were scattered throughout the Bekaa Valley, Lebanon. Monzer al-Kassar, a Syria arms and drug smuggler was associated with the chief of the Syrian intelligence by marriage. Al-Kassar was identified as a key player in the release of the American hostages, as he had orchestrated the release of the French. In return, he wanted protection of his drug routes.

A deal was made with a rogue group of CIA operatives working out of Germany and known as CIA-1. The CIA-1 agents were planning a similar deal to trade drug shipments for the release of the hostages. The operation was known as COREA. The shipments of drugs were then sold to the drug kingpins in the USA, thereby penetrating their distribution systems. The idea was to prepare the ground for future drug busts. The profits, meanwhile, went to fund the Contras.

The drugs were being sent to the USA via Frankfurt and London in what was known as Controlled Deliveries. The German, UK and US Customs had been told to let the bags through for "national security purposes." Pan Am was one US carrier utilized for such shipments — and without notifying Pan Am's management of these covert operations. These shipments were on regular commercial flights with regular passengers on board.

One of the carriers, or mules, was a young Lebanese-American named Khalid Jafaar. His death onboard Pan Am 103 raises more questions. Jaffar was known to the various customs officers and had made a number of "controlled deliveries" without incident. The fact that he had the freedom to pass through customs without being searched was well known and it is possible that one of his suitcases, loaded with drugs, was switched at Frankfurt or Heathrow. It has been suggested that the switch might even have been the work of the CIA, whose COREA operation was about to be exposed to Congress by McKee.

As the story goes, when McKee found out that drugs were being shipped to the USA in a drugs-for-hostages operation, he was not about to sit still for it. This rogue CIA-1 "COREA" operation was by now operating brashly and recklessly under his nose.

Both McKee and Gannon died on Pan Am 103, despite warning of a potential bomb — in the famous Helsinki phone call warning of the bombing, and the notification on the US embassy bulletin board in Moscow. The word certainly got around, but apparently it did not get to McKee and Gannon. Curiously enough, five people on the McKee rescue team were booked to go on Pan Am 103 — a few of them changed their flight plans, and the question remains, why?

The Helicopter. Within one hour of the crash, the whole area around Lockerbie was swarming with Americans, some with FBI jacket and caps according to David Ben-Aryeah, a local journalist (who claims to have the most complete file on the Lockerbie crash and who eloquently narrated the documentary, *The Maltese Double Cross*). Given that this event occurred over Lockerbie, the local Scottish police had sole jurisdiction over the collection of evidence and the reporting thereof.

Initially, a white, unmarked helicopter was seen coming and going, all over the crash site, with FBI agents moving bodies, and removing tags that the Scottish police had placed on the bodies. A brown Samsonite suitcase was removed; according to a local farmer, it was loaded with American money, and another was found full of drugs. Later, the suitcases were returned to the same spot.

Under Scottish law, the police are forbidden to discuss or disclose any information about the investigation, but local people, employed in the recovery of the debris, were quite willing to relate the facts to the media. The stories of the brown, or bronze-colored, Samsonite suitcase will always be a difficult aspect of the Pan Am 103 trial. Just a month or so before the Lockerbie trial, the Scottish police admitted that a considerable volume of their investigative work had been lost! How could this happen?

Helicopters, some military, were constantly buzzing around the site for over 48 hours, dropping off people who were in a frantic search for something. This is not normal at a crash site.

The Simulated Tests: Pan Am 103's Container 4041 Blown Up. Boeing 747 Blown Up; Libyan-Russian Plane Blown Up. There were probably dozens of simulated explosions conducted as tests during the Pan Am 103 investigation. Later, we would also be reading about the experiment simulating the TWA 800 central tank explosion. For now, let us concentrate on three tests conducted in the Pan Am 103 investigation.

When the bomb went off in the brown Samsonite suitcase in Pan Am 103's container 4041, which had been located near the side of the Boeing 747, it supposedly blew out a hole, 17 feet in diameter, in the side of the fuselage. Yet other similar bombings had only created holes of about 2-3 feet in diameter. Luggage containers are made of aluminum, with the minimum strength to do the job; weight must be kept as low as possible. The Pan Am 103 container flew out of the hole and dropped 31,000 feet to the ground. Yet pictures we have seen of this container show that it had clearly survived the crushing force of this high-impact landing, after dropping at a speed that would have exceeded 100 miles per hour.

The next simulation is the most troubling. The US investigative team, either the FBI or the NTSB, acquired a retired Boeing 747-100 and placed a bomb in it. The whole episode was videoed for, I assume, the benefit of both investigators and the general public. It was very vivid and, in some ways, unnerving. It must have been very dramatic for the victim's families, who could now imagine the suffering that their relatives and friends must have experienced just before their deaths.

The problem I have with the simulation is that the bomb was placed just behind the wings and not up-against the bulkhead, where Container 4041 had been placed. Why? And did they place the test bomb in a Toshiba 453 radio, in a suitcase, in a container, to accurately simulate the explosion? Had this Boeing 747 been pressurized to simulate the conditions at 31,000 feet?

Imagine the power of the four jet engines as they lift a 350-ton Boeing 747 off the runway, its nose high in the air. (It's quite a sight.) The strongest part of the plane is *forward* of the wings. The fuselage and tail section are nowhere near as strong as the rest of the plane — they do not have to be. Yes, they are strong; but only strong enough to be dragged along with passengers, freight and the day's meals. Placing the simulation bomb behind the wings, in the weakest part of the plane, was certainly done intentionally, for maximum publicity. The bomb did not just blow a 17-foot hole in the side of the plane, it demolished the back end. The tail section crashed to the ground and the front of the plane looked as though nothing had happened (it's true that in real life, the front nose wheel would have had a lot more weight). The video was shown over and over again on CNN.

Now, had the simulation bomb been placed in the same position as container 4041 in the hold of Pan Am 103, the outcome should have been quite different. If the investigators' theory were right, there should have been a 17-foot hole in the plane, which would otherwise have remained intact. The force of the headwind facing Pan Am 103 could then have entered the hole in the side and cause the breakup of the remainder of the plane.

The fact is that in a different episode, a United Airlines Boeing 747, with an even bigger hole in the side, managed to land. This appears to have been ignored in the investigation.

It would be most interesting to witness another simulated bombing on the appropriate vintage Boeing 747-100 plane, with a similar maintenance program, but this time with the bomb inside a radio, in a container, in the same place as Container 4041 on Pan Am 103. I suspect the result would be a 2-3-foot hole in the side, with Section 41 falling to the ground. That would explain why the Section 41 landed 4.5 miles closer to Heathrow Airport than the rest of the debris, and would be a more accurate account of events.

Kuffra lies in the southeastern area of the Libyan dessert, not far from the Sudan border. I knew the area well, since my neighbor in Tripoli ran OXY's Kuffra project. The Libyan desert is underlain by a huge aquifer containing enough fresh water to feed the Nile River for 2,000 years. Where this aquifer approaches the surface, it forms oases. OXY, under the terms of its oil production agreement, designed and built a system of shallow water wells and sprayed the water on alfalfa crops. The alfalfa was fed to sheep which, when mature, were sent to the Tripoli market. That was back in the late 1960's. Later, Libya had an airstrip built in the area. That is where the next test took place, conducted by the Libyans.

To witness a simulation of a bomb detonating at altitude, Bollier flew to Kuffra (in a Russian-made plane similar to the American C130, owned by the Libyan military). The test plane was rigged as closely as possible to the Pan Am 103 Boeing 747 and sent aloft by remote control. Bollier told me that there are three microphones on a Boeing 747, each so sensitive that it can pick up the noise of a small hammer hitting the side of the plane at any point on the fuselage. On the Libyan test, the noise of the explosion was heard briefly on the CVR, the Cockpit Voice Recorder, and he wonder why it had not been detected on the CVR on Pan Am 103.

There was no telltale noise on the CVR of Pan Am 103, nor of TWA 800 or Air India 182. This coincidence has led many investigators to make similar claims. Remember the Boeing insider, who wrote to Bollier to confirm his "no bomb" theory. The simple explanation would be that Section 41 separates from Section 42 while in flight, cutting off power to the Black Boxes, the Cockpit Voice Recorder and the Flight Data Recorder.

Bollier and I discussed this at length and, since he strongly believes there was a bomb, I am going to accept the fact that a bomb did, in fact, trigger the demise of Pan Am 103. I also believe there was a rapid chain of events. The bomb went off, driving a two-foot hole in the side of Pan Am 103; Section 41 separated from Section 42, and the force of the onrushing air enlarged the hole in the side of the plane; containers flew out into the cold dark air at 31,000 feet. This all happened in a split second.

Pan Am 103, a Boeing 747-121, registration number N739PA, was the fifteenth model rolled off the Boeing assembly line in

Everett, Washington. Companies such as British Airways, which demanded expensive modification before it would purchase the planes, did not consider this plane airworthy as it was first presented. The Boeing 747 had fatigue cracking problems. The plane in question was owned by Pan American Airways, which on the verge of bankruptcy already and may have ignored problems with its fleet of older Boeing 747 aircraft.

Bollier Set Up As Fhimah? The CIA had been watching Bollier and his business excursions into Libya from the get-go. They knew exactly what he was selling and to whom. They followed his every move, as he was manufacturing equipment for the Libyan military (albeit with the approval of the Swiss authorities). How many millions of people are spying on the worldwide business community?

On one of Bollier's excursions in 1988, he was booked to go from Tripoli to Zurich via Malta, but at the last minute he managed to get on a non-stop Swiss Air flight to Zurich. He did not cancel his flight to Malta or cash in his ticket. He was subsequently told that he had accompanied the famous brown suitcase on the flight to Malta, for Fhimah, one of the alleged Pan Am 103 bombers.

Last Minute Cancellations on Pan Am 103. Pan Am flight 103 was fully booked, a few days before taking off on December 21, 1988. The youngest daughter of Dr. Jim and Jane Swires could not get on the flight. Then, at the last minute, there were so many cancellations that her booking was accepted, along with those of many other young people going home for Christmas.

President Botha of South Africa and a number of his military aides had switched to an earlier flight and his aides, including military officers, safely went back to South Africa. Had they been warned of potential trouble?

When Gannon and McKee decided to go back to the USA early, McKee took the Sunni boat from Lebanon to Cyprus, and Gannon and the American Ambassador, Jack McCarthy, flew in by helicopter. McKee, Gannon, O'Connor and LaRiviere then flew on to London and boarded Pan Am 103; the Ambassador decided not to go.

The chief investigator on the Lockerbie case, Oliver "Buck" Revell, had a son stationed in Germany who was booked on Pan Am 103, then delayed his trip. There were other government employees who also changed their reservations at the last minute.

One very important part of any aircraft crash investigation is to check the identification not only of those who took the flight but of those who changed their reservations. One Samir Mohamed Ferrat, for example, died on Swiss Air 111 on September 2, 1998, off Peggy's Cove, Nova Scotia. Ferrat was doing business in Africa with Ron Brown, the Secretary of Commerce in the Clinton administration. Brown, you will remember, was killed in a plane crash over Bosnia. The documentary covering this story was shown on Swiss TV during its investigation of the Swiss Air 111 crash, but the possible linkage was never followed considered by the British and American media. The Swiss believe a microwave device known as a HERF gun brought down the plane. This is in *Information Warfare*, by Winn Schwartau. The key to this story was the fact that Ferrat had received hundreds of millions of dollars for housing in Africa and the US taxpayer's money had disappeared!

First Syria and Iran were held responsible for the bombing of Pan Am 103, then the focus shifted to Libya. One documentary produced in the US by *Frontline* was called, "The Bombing of Pan Am 103", it was aired on January 23, 1990, some two years before Libya was accused. *Frontline* is a respected TV investigative program, and its reports are usually accepted as well-researched, accurate and honest. Much of the footage from this program was re-used in subsequent investigative documentaries in both the US and Britain. The focus was clearly directed to the possibility of a bombing masterminded by Ahmed Jibril of Syria, at the request of Iran. Iran was seeking revenge for the shooting down of its Airbus in the Gulf.

Early in the program, the Scottish police hold a press conference and announce that a bomb brought down the Pan Am 103, based on preliminary analysis of bomb residue on a piece of Container 4041. But, as Bollier has pointed out, the picture of Container 4041 did not look as though it could have fallen from 31,000 feet. The question becomes, what container part were the Scottish police referring to?

Another important point the documentary sought to convey was the total indifference of Pan Am's management and all of the government agencies. The victims' relatives were very vocal in their criticism, not only with respect to their grief but also to the cavalier attitude towards the facts that might indicate who was responsible for the crash.

The attitude of the investigators was such that the relatives got together and appointed group members to represent them. Bert Ammerman was appointed for the US victims and Dr. Jim Swire for the British and Scottish. These two men quickly developed a very jaundiced eye towards the investigation; each ended up conducting his own research. The two men traveled extensively, interviewed many people, studied the evidence and concluded that the story was by no means as cut and dried as the authorities were trying to portray. Both Swire and Ammerman stated publicly that the two governments were hiding the truth. They also stated that they doubted the truth will ever come out.

Frontline made it quite clear that evidence pointed to Jibril, and to a Palestinian named Mohammed Abu Talb. Talb was living with his family in Sweden during this period and was a known terrorist for the PFLP. The story goes that he had been followed on a trip to Cyprus, Malta, and then to Frankfurt, where he met with Dalkamoni. In Malta, he purchased men's and babies clothes, apparently at random, according to *Frontline* (who interviewed the owners of the shop, Mary's House). Parts of the MEBO timer, we understand, were found in the burned fragments of baby clothes.

Much emphasis was placed on the random selection of the clothes, yet there could have been a simple explanation. Perhaps Talb intended to buy clothes for himself and his baby, but when he met Dalkamoni, they decided to place the Toshiba bomb in the same suitcase. Talb is currently serving a life sentence in a Swedish jail for other terrorist events (based on the flimsiest of information, according to David Ben-Aryeah). The whole bombing chronology was clearly presented and it directed the focus to Jibril. Libya was only briefly mentioned, in this program, as a state funding the PFLP.

The *Frontline* documentary shows a hearing before the US congress, where various experts present their theories and recommendations, particularly with respect to potential future attacks. One expert discusses the various ways in which bombmakers can disguise their deadly wares. I was amazed the see a whisky bottle bomb, which the expert said was found on KAL 007. How did they find that bomb, in the Sea of Japan, when they could not find Section 41, with the cockpit and the cameras?

The narrator then discusses the bombs found in Dalkamoni's car when the BKA, the German police, made their "Autumn Leaves" arrests. The story goes that four radio bombs were found and a fifth was missing. I am not sure how they knew one was missing. What was very obvious and relevant was the picture of the Toshiba Radio-Cassette Player. It was the Toshiba Bombeat 453, measuring just 10 x 7 x 2 inches, with one speaker. The Toshiba radio story was later switched to refer to a two-speaker version, the Toshiba Bombeat 8016. (The Toshiba radio shown on *Frontline* was not a "Bombeat", by the way.) In the *Frontline* version of the evidence, produced before Libya was named as culprit, it was quite clear that the radio was the one-speaker model.

In *The Fall of Pan Am 103 — Inside the Lockerbie Investigation*, Steven Emerson and Brian Duffy describe in detail how cleverly the Toshiba Bombeat radio-cassette player had been packed with the Semtex-H explosive! Later in the *Frontline* program, the narrator tells us that the trigger mechanism was a barometric device. How did they know it was a barometric trigger? If that is true, then bomb must have been placed on board Pan Am 103 at Heathrow. Burt Ammerman told a group of people, shown on *Frontline*, that the British security people who attended a meeting at Heathrow were downright hostile. Did they know something that we all need to know? Did they tell Burt that the Iranian baggage handling system was next to the Pan Am facility?

Almost at the end of the program, there is a section in which some of the relatives meet with President Bush at the White House. Bush makes it clear that he will do everything in his power to find out who bombed Pan Am 103 and that he would most certainly retaliate. One of the relatives then spoke to *Frontline*, saying, "We shall see what happens. We shall see if Bush retaliates — then I will judge the man."

Chapter 7

TWA 800 AND THE FOUR CRASH THEORIES

There could be many reasons why TWA 800 fell from the skies in the evening of July 17, 1996. The authorities chose three and pursued them with every means at their disposal. The fact that a fourth reason may have existed was, for the most part, ignored. The three were: international terrorism, a missile (either from the United States Navy or an enemy), and an electrical discharge that would have caused the central fuel tank to explode.

I will attempt to describe why each theory was worth pursuing and will review some of the hypotheses that the various news channels, talk shows and papers promoted on a daily basis. Some were serious, many were laughable, and a few were bordering on criminal, in their own right, since they hampered the effectiveness of those who were legally in charge of the investigation.

The Terrorism Theory

There is little doubt that terrorism was, in many people's minds, the most likely explanation — even though terrorist acts in the past had mainly been perpetrated by groups of people who hijacked the planes and only threatened to blow them up. In many cases, the hijackers were brought to justice without death and and/or damage to the equipment.

In the case of Pan Am 103, it is alleged that the bomb was placed on the plane but that the terrorists did not accompany the bomb, even though there have been dozens of cases where individuals have been prepared to die for their causes. Many nationalities have, indeed, engaged in terrorist acts involving airplanes, and disgruntled Arabs have been the most visible of them. Their cause has, for the most part, been linked to their grievances with the Israelis over the Palestinian problem. As a consequence, the first reaction is to blame the Arabs. To be fair, one could expect such a reaction from certain Arab factions, based on the allegations against certain Arabs in connection with the World Trade Center bombing in New York. Some groups consider themselves at war with the West and with the US in particular, for its continuing support of their enemy, Israel.

On November 13, 1997, ABC's Peter Jennings announced that the FBI had completed its investigation of the criminal element of the TWA 800 crash, after an exhaustive search for the truth, which included interviewing 7,000 people.

The FBI concluded that the plane must have suffered a mechanical failure caused by an explosion in the central fuel tank.

TWA 800 and the Four Crash Theories

The animated analysis of what happened to the Boeing 747 was very well done, very graphic, and it coincided exactly with the reasoning in this book.

The Missile Theory

One of the most distracting theories held that the aircraft was accidentally hit by a test missile by the US Navy or the Air Force. This was a very interesting exercise, which can be placed in the same category as the idea that the plane was hit by a meteor or falling debris from space. It's certainly bad luck on the part of TWA if Flight 800 was hit by space debris but, if you lack any other plausible explanation, it is certainly not impossible, however remote.

The missile theory, I believe, would have died quickly had the US Navy and Air Force come out quickly with a response. Neither military force said a word, which simply fueled an already biased public into calling this yet another covered-up conspiracy. Both forces must have known immediately that this event was not of their making, yet they gave the appearance of seeking to avoid questions while they struggled to come up with an appropriate response. Surely, with hundreds of billions of dollars worth of satellites, and thousands of radar units monitoring the US coast, they knew very well whether any missile had been accidentally fired or whether a terrorist was shooting planes out of the sky above Long Island!

The missile theory was really hatched on the basis of the observations and photographs of some residents of Long Island, near Moriches Inlet. I first saw the pictures while I was strolling down the main street of Geneva, Boulevard Mont Blanc, towards the lake. I just happened to see a copy of the November 7, 1996 *Paris Match* magazine, and the caption *TWA 800* caught my eye. The French thought they were onto something big. I was amazed at what I saw. The picture of the alleged missile showed it streaking towards the ill-fated TWA 800, with the white-to-red exhaust trail in front of the missile. That's strange, I thought, isn't the hot end supposed to be at the rear of a projectile? At first, I considered whether the picture could be showing the missile coming down, having already hit the plane. That would be one explanation; but why was there no ball of fire, with the plane blowing to pieces, given it had just taken off and had a full complement of jet fuel? In fact, the missile should have been destroyed on impact; that is its mission in life. On the other hand, the missile could have had a dummy warhead, for practice, and it could have passed straight through the fuselage and remained intact. An unlikely scenario.

Pierre Salinger's explanation was elaborate, with pictures and diagrams; he was quite frank about his conviction that this was a government conspiracy and cover-up. The media had a field day, on the basis of his analysis, but nobody who took the trouble to think it through could have believed his conjectures. Franck Curtin told me that Salinger had been paid $5 million to introduce this forceful analysis into the public discussion, in order to throw people off the scent. Under no circumstances could we afford to let people suspect mechanical failure; it would have caused panic. This tale may be apocryphal, but Pierre is even rumored to have boasted about his reward to one of his neighbors in Paris. Regardless of any truth to the tale, his theory certainly had the desired effect.

It seems conceivable that Salinger was, indeed, paid to put this theory into circulation. He and his erstwhile friend, Ian Goddard, announced on November 6, 1997, that the whole episode had been fabricated. Strange that this should happen just seven days before the FBI announced that it had been unable to attribute the TWA 800 crash to terrorism of any form.

The best analysis of the missile theory, based on a technical understanding of flight, is in *The Downing of TWA Flight 800. The Shocking Truth Behind the Worst Airplane Disaster in US History*, by James Sanders. The author is certainly skilled in math and physics, and offers a credible scenario of the events that fateful night, although some of the assumptions seem off-base.

Sanders painstakingly plotted the position of each of the main parts of the wreckage and by careful trigonometry calculated the position of the parts as they separated from the plane. He shows that Section 41 was some distance from the remainder of the plane debris. He failed, however, to mention the fact that the door on the nose wheel was the closest piece of debris to the departure point. This was clearly the first piece to come off, and the FBI has chosen to ignore this important fact. The nose wheel door information was published on CNN's webpage on November 14, 1997. Too bad that Sanders did not have the "Boeing Report"; he would have been way ahead of the game.

The Central Fuel Tank Explosion

The NTSB (US National Transportation Safety Board) had several options to consider under the central fuel tank explosion theory. They soon decided that the most likely scenario was that the central fuel tank exploded, causing the nose of the plane to fly off first, then the left wing, followed by the remaining parts of the plane.

The Boeing 747-100 has the capacity to fly far further than from New York to Paris; but in the interests of economy and sound business practices, a plane is only loaded with enough fuel for the trip (with, of course, a margin for safety, in case the plane is diverted or kept in a holding pattern before landing). The amount of fuel provided as the safety margin is mandated by law, although a pilot will have discretion to load more than the required amount if he suspects he may need it.

Under normal operating conditions, the mandated amount is equal to the quantity of fuel required to keep the plane flying for one hour after it arrives at its destination. We can assume that TWA Flight 800 had enough capacity in the wing tanks for the trip and the central tank was left close to empty. This distribution of weight has apparently been found to be most desirable.

I will review each of the fuel tank theories in some detail. First, there was the theory that the fuel became overheated. The air-conditioning units were located directly under the central fuel tank, which could have caused the fuel to vaporize, thus increasing the pressure inside the tank. There was very little fuel in the tank, but let us assume that the pressure did, in fact, increase inside the tank. Certainly, the design engineers would have been very well aware of such possibilities and the tank would have been designed with considerable extra strength in order to ensure that the tank would not rupture, while in flight, with a fully-loaded tank. These planes are designed to travel in very turbulent weather and in fact can take more punishment than the passengers could sustain. Imagine the force of the fuel, sloshing around in stormy weather. . . and the tanks are built to withstand that, and more.

If this scenario had any viability whatsoever, then all planes would have been grounded since they are all designed, for the most part, with the same systems. Thousands of planes are in the air at any one time, and none has experienced this type of tank failure. Still, let us imagine that the pressure was abnormally high, to the point where it might burst the central fuel tank. Boeing's engineers would have installed a pressure-release value in the system, to vent excess pressure to the atmosphere. (It is probably located in the fuel dump system.)

Then we have the hypothesis that an electrical wire shorted out in the fuel tank and caused an explosion. This became one of the most popular theories, even though (or because?) it is one of the least credible. No plane has ever been built with an electric wire passing through any fuel tank. No designer, at any level of training, would consider such arrangement, for obvious reasons. Apart from the sheer stupidity of taking any risk with such clearly dangerous elements, trying to make sure that the wires were in an explosion-proof casing would be an assembly nightmare and would add unnecessary weigh and complexity to the overall design. That dog just won't hunt, as they say.

Still, let us suppose that this old fuel tank had developed a small leak (near the bottom of the tank, since the tank was almost empty). Now, we have fuel outside the tank, where an electrical circuit could conceivably have been. The pilot was climbing, at the time of the accident, and he easily could have thrown a switch (to dim the cabin lights, for example), causing a short circuit. If the electrical cable was sparking, it could cause the fuel to ignite — but *outside* the fuel tank.

There have been many cases where fuel leaks have occurred outside the tanks. The B-1 Bomber had just such a problem. The flaw in this reasoning is that fuel does not explode as readily as one might think. The conditions need to be exactly right.

Moviemakers love major explosions, as the spectacles adds to the drama of the story. But the explosives they use are rarely gasoline or jet fuel in a tank. These explosions are caused by a well-orchestrated ignition of a meticulously prepared fuel-and-air mixture. Another example is the so-called explosion that occurs in the pistons of a car engine. While the fuel does burn at a very rapid rate, these are not explosions, *per se*. The fuel injected into the pistons is very precisely measured to coincide with the requisite amount of air. When the spark plug triggers, the fuel is ignited and a "burning front" progresses in the cylinder in a controlled fashion, albeit at a very rapid rate. In fact, explosions can occur in car pistons and the designers are careful to avoid such events, as they would slowly destroy the cylinders and pistons. For any fuel to explode spontaneously, the system needs to have a precise amount of air and fuel and the mixture must be homogeneous. The mixture cannot explode if there is too much, or too little, fuel for the air. There is what is known as an "explosion envelop" in which the mixture is right to cause an explosion. It is fairly simple to cause fuel to burn, but not to explode.

TWA 800, according to the fuel tank theory, was destroyed by an explosion. If the central tank (which is located near the bottom of the plane) exploded, why did the nose, or Section 41, fly off first, just like the Pan Am 103 Boeing 747-100?

C-Span broadcast a video of a meeting of the "Accuracy in the Media" group. The speaker clearly understood the theory of explosive mixtures and showed the audience a home video of an experiment he had conducted in his backyard. He had mounted a metal dish on his barbecue and warmed up some Jet Fuel A, which would have been the same fuel used in the tanks of TWA 800. He then lit several matches and made any number of attempts light the fuel (nevermind getting it to explode). Jet fuel is much heavier than gasoline and it is difficult to light. In a jet engine, the air-fuel mixture is carefully controlled.

The NTSB built a one-quarter sized central fuel tank model to conduct a test of what might have happened. It was a farce. The tank had a small amount of fuel inside, with a sparking device inside the tank. The folks conducting the test were singularly unsuccessful, even though they tried many combinations. I heard that they then introduced a small quantity of propane — and bingo! the right side of the tank blew out. Even from the color of the flame, it can be seen that the burning fuel was not simply Jet Fuel A! Still, CNN played this mock tank explosion over and over in the US.

Another documentary about TWA 800 clearly was made to placate the traveling public. An NTSB representative shows a reporter how the exploding fuel tank had forced the front section of the tank forward, so that the top of the tank hit the bulkhead. The marks were quite clear and the explanation perfectly reasonable.

Let me bring in the wisdom of one more outside expert. There are a number of people around the world who look back with pleasure and pride on the amazing days of the de Havilland aircraft company. People tend to forget that the De Haviland Spitfire, Mosquito, and other famous World War II fighters where instrumental in preventing this book from being written in German. One such person is Ron Davies (no relationship to the author), Curator of Air Transport at the National Air and Space Museum, Smithsonian Institute, Washington, DC. Ron is an expert on worldwide commercial transport airplanes and has written many authoritative books on the subject. He summed up his view of all the hypotheses surrounding the "central tank fuel tank explosion" theory in one word: "Nonsense"

No sound explanation has been presented to support the exploding fuel tank theory. And, if that were a real problem, every plane should be grounded for safety reasons.

The Metal Fatigue Theory

If you take a metal clothes hanger and start bending it over and back, it will soon break. Hangers are made of cheap metal, to serve a simple purpose. Bending the hanger causes metal fatigue, and at some number of stresses, the metal fails. The technology of metal fatigue is well understood, and designers can make products with a certain specified life expectancy. The industry calls this "designed obsolesce". Equipment is often guaranteed for a fixed period of time, during which the manufacturer is quite confident his product will be unlikely to fail. (The Japanese made great inroads into the American car market by designing cars with a much greater effective life expectancy and low maintenance cycles. Children's toys are notoriously short-lived, often lasting until only a month after Christmas, when the manufacturers know the kids will be tired of the toy anyway and will want something new).

The metal in airplanes undergoes this analysis, but with a far greater attention to the technical details. Airframes are usually manufactured for a twenty-year life cycle, and many planes are still airworthy even forty years later. However, modern aircraft are designed to much tighter tolerances based on economics, greater confidence in technology, and the market for the products.

As I mentioned earlier, the de Havilland Comet 1 suffered from metal fatigue around the square windows; on the other hand, we rarely hear of planes crashing due to metal failure in a well-maintained aircraft. It does happen, but rarely.

One recent glaring example occurred on April 28, 1988, when Aloha Airlines Flight 243 (a Boeing 737-200, N73711) lost a large section of the roof, just behind the equivalent of what would be Section 41 on a Boeing 747. The accident occurred near

Maui, Hawaii, at about 24,000 feet, while the flight was on its way from the Hilo airport to Honolulu. When the roof came off, the plane immediately decompressed and a stewardess was unfortunately sucked out of the plane. The plane was brilliantly controlled by the pilot, who landed without further loss of life. This Aloha Airline plane clearly suffered metal fatigue — a case of poorly maintained equipment.

Chapter 8

WHY DO BOEING 747'S BREAK UP IN FLIGHT?

The simple answer is metal fatigue and failure. And what causes the fatigue and ultimate failure of certain critical parts? Before going into the specifics of what happens to early Boeing 747's when these problems manifest to the point of complete failure, I am going to describe briefly the forces that are at work on all commercial aircraft, both on the ground and in the air. This will explain clearly what actually happens.

Stress on All Sides

Visualize any modern plane, as it sits on the apron at the airport or as it is taking off or landing. The main wheels are generally under the wings, and there may be as many as eight or more wheels on each side. These wheels are known as the undercarriage; they bear almost all of the weight of the plane. When a plane lands on the runway, it comes in "nose high" and it lands on the main wheels. The nose is kept in the air until the plane has almost completed the landing. Only then does the pilot gently lower the nose onto the nose wheel (steering wheel).

The nose wheel usually has two tires and is considerable lighter in structure than the main wheels. The plane is designed to be slightly nose-heavy, to balance the plane and give the pilot control over the aircraft while it is on the ground. While the plane is standing on the apron being loaded with passengers, its own form creates various stresses. The weight of the fuselage and tail, behind the main wheels, tends to bend the fuselage downward. In front of the main wheels, the forces are more complex, since the weight of the tail acts to lift up the nose while the weight of the nose keeps it in place. Furthermore, the part of the fuselage between the main wheels and the nose wheel is inclined to sag (to a degree which is not noticeable to the naked eye, but the stresses are at work nonetheless).

Now, during take-off, the plane rolls down the runway and the pilot pulls back on the yoke; the nose of the plane leaves the ground. The weight of the plane is again on the main undercarriage. The weight of the plane's nose bends the fuselage downward, in front of the main wheels, and the tail pulls it down behind the wheels. As the plane leaves the ground, the forces are now totally centered on the wings and the plane does what it does best: fly. The same forces are at work when the plane lands.

During flight, of course the forces are far more complex, including headwind, prevailing winds, turbulence, and so forth. In rough weather, the body of a Boeing 747 can waggle like a tadpole in rough water. This is quite normal, and if the plane did not flex, it would break up very quickly; or it would have to be so strong and heavy that it could not fly anyway.

Now consider those long, thin, sausage-like balloons that can be twisted into animal shapes. When you begin blowing, the balloon is sagging, but after a few puffs it begins to straighten out. Once it is fully charged, the balloon becomes relatively strong and hard to bend. Similarly, airplane cabins are pressurized from within. When you travel in a commercial airplane, the cabin pressure is maintained — not at ground level, as many people believe, but at the pressure one would experience at 7,500 feet. At this pressure the passengers and crew feel no ill effects from the altitude and therefore there is no reason to increase the pressure. When the plane descends below 7,500 feet, one often has to adjust the pressure inside one's ears.

If passengers were to insist on sea-level pressures during flight, the planes would have to be built much stronger to withstand the difference in pressure between the inside of the plane and the rarefied air at, say, 35,000 feet. Two factors come into play when the plane is pressurized. One, the plane actually gains in rigidity, like the balloon described above, while, two, at the same time, the metal fuselage is placed under a modest stress. A small amount of air, under a small pressure, makes so much noise when a balloon bursts...

Chickens lay round eggs for some very good reasons, and the hen's comfort is not the only one. If the eggs were rectangular, they would break extremely easily. This fact has not gone unnoticed by the designers of aircraft, which is why most plane bodies are circular. When the plane is pressurized at altitude, the pressure is equally distributed in the fuselage so that no one area is stressed more than another.

It is important to note that the Boeing 747 was probably first commercial plane to have an oval cross-section in the front part of the fuselage. The plane has been designed to take into consideration the effects of pressurization, and normally there will be no problem. Take your sausage balloon in its inflated state, and with your two hands press the front part of the balloon into the oval shape of the Boeing 747. It requires a surprising amount of force. Take your hands off the balloon and it quickly assumes its original circular shape. These forces are at work on the Boeing 747 when it is flying at altitude. The sides of the plane naturally seeks to become circular, placing outward force on the central part of the fuselage that is designed to keep the plane in its required shape. This outward force, trying to make the Boeing 747 fuselage circular, is a very important factor in our analysis since it is clearly an area under considerable stress that does not exist in any other commercial plane.

One more force is at work. When a large plane such as the Boeing 747 is flying, almost the entire weight of the plane and its contents is supported by the wings. However, as the plane reaches its cruising speed of approximately 550 mile per hour, a certain amount of lift is generated by the belly of the plane, particularly in the front section, around Section 41. This effect can best be described by the analogy of a ski on water. The force of the air going under the belly of the plane adds lift, but at the same time it places a disproportionate upward force on the nose of the plane. At 550 MPH this force is substantial and it is compounded by the forces on the nose of the plane as it forces itself through the air.

The frontal profile of the Boeing 747 is very large and the forces exerted on Section 41 as it travels at cruising speed are massive; it takes a very clever design to handle all these factors. If you hold your hand out of the car window at only 60 MPH, you can feel not only a significant backward pressure but, if you hold your hand like a wing, you will feel the upward pressure as well; you may be surprised at the force. During a hurricane, houses and cars are tossed through the air by winds of a mere 150 MPH! Yet, even at 550 MPH, airplanes are safe — as long as the outer skin stays intact.

Skin Problems

The problem with early Boeing 747's is the quality of the aluminum used in its manufacture. In the early to mid-1980's, Boeing began to discover an unusually high incidence of structural fatigue and failure in Section 41 — well before they were anticipated, according to the original design.

These stress fractures were appearing regularly, in various locations but in greater concentration in Section 41. This was, of course, a matter of grave concern to both Boeing and the FAA. Boeing and the FAA mapped out a program to conduct the repairs necessary to keep the planes flying safely, and they distributed the work program to many qualified repair centers around the world, including Franck Curtin's potential operation in Arizona.

When a plane undergoes such massive required repairs,

costs include not only the repairs but the income that is lost while the plane is on the ground. Airlines are reluctant to take planes off-line and in the case of TWA, they simply could not afford the repairs; they would be forced out of business. In fact, in 1996, TWA had the oldest fleet of planes in the business. They have been in bankruptcy many times and have had trouble meeting their financing obligations. This was especially true in the 1980's and early 1990's, before the airlines began making so much money in the mid-1990's.

According to Franck Curtin, Boeing 747's (numbers 15 through 686) were made with aluminum produced in the (then) USSR. The aluminum was an inferior grade, different from what had been specified, and Boeing did not notice the problem until it was too late.

There were simply too many planes in the air and to have them taken off-line would have caused chaos in the worldwide air transport industry. It might easily have caused a worldwide recession, or even a depression. Was Boeing more interested in landing new orders and competing with the European Airbus, at that time, than in diverting its resources to correcting this problem? What about Boeing's prestige?

A Little Help from Our Russian Friends

I had to check out Mr. Curtin's allegations. How could I get information on Russian aluminum production and the plants that may have produced the parts for the Boeing planes? This data was not readily found on the Internet. I searched the libraries in London and found a certain amount of information, which clearly demonstrated the pathetic situation of the Russian aluminum production in general. The plants were very old, were highly polluting, totally inefficient and clearly uneconomic by world standards. However, I could not determine exactly which plants had provide the aircraft parts.

Then I remembered that many Russian planes had crashed due to "mechanical failure"; poor quality aluminum was a very reasonable explanation. In my notes, I found a particular reference to such an incident. In the *US Today — World Edition* dated March 31, 1997, was an article entitled "Rust may have caused Russian plane crash". The article states that the fuselage was so rusty that it fell apart in the air. Of course, planes are not made of iron, so "corrosion of aluminum" might be more accurate than "rust", but the writer's point is that the Russians were not producing high quality aircraft parts. Actually, many years ago a Russian pilot defected to the West with a Russian jet fighter and the US engineers were surprised to find a few parts made of steel, which could indeed have rusted. The article goes on to say, "The Antonov AN-24 twin-turboprop airliner, operated by Stavropol Airlines, was flying at 19,700 feet when it broke into pieces and crashed near Cherkessk." So there is a history of Russian planes breaking up in flight, like a few of the Boeing 747's.

Ron Davies had told me about the brilliance of Russian aeronautical engineers — although I have noticed that many of the Russian planes bear a remarkable resemblance to US-made planes. However, under the Russian system, I am sure the aeronautical engineers knew that the aluminum was inferior but could say nothing.

More proof and support for the Curtin allegations was not hard to find. *Giant JetLiners* (Guy Norris and Mark Wagner) has a section on "Pressure and Fatigue" that is loaded with facts and observations that clearly support Mr. Curtin's first-hand knowledge. Boeing had recognized that serious and unexpected fatigue cracks were appearing in Section 41 long before the design team had anticipated them. On page 108, Norris and Wagner discuss the metal problems and I quote, "Boeing started to use a more fatigue-resistant aluminum copper alloy in the frames around the nose, instead of the original aluminum-zinc alloy. The manufacturer also made the same changes to the older 747's in service, so they required less frequent inspections".

The question becomes, where did the original aluminum come from, for at least some of the main parts? The subject of where these parts came from will be brought up again in my discussions with Mr. Bollier, owner of MEBO.

Fail-Safing Successes

On page 102 of *Giant Jetliners*, the authors confirm that the techniques of aircraft manufacture had changed little since the building of the de Havilland Comet in 1949. Of course, metal technology and manufacturing organization have improved immensely, enabling a huge leap in scale of manufacture. They add, "Like many generations of airliners before them, each was designed with a fail-safe structure. This meant that if a part of the airframe should fail for any reason the stresses and loads imposed on that part of the aircraft would be carried by other parts and diverted around the rupture by different *load-paths*".

Boeing has stated many times that the 747's can sustain a suitcase bomb and still land safely, with a minimum loss of life. There have been explosions on Boeing 747 aircraft and they have landed, and been repaired; a few examples were cited in the Introduction.

On page 107, and again I quote, "With the alarming increase in terrorism and bomb threats in the late 1979's and 1980's, a series of new specifications were drawn up by European and US aviation officials. *These specifications stipulated that all new wide-bodied designs must be able to sustain a 20-square-foot hole blown in the side of the fuselage without endangering the floor and essential control runs.* One aircraft, the Boeing 747-400, (the newest models) was briefly brought under scrutiny by this because only the main deck could meet this criteria; the upper deck was only able to sustain a sudden blowout producing a 12.5 square-foot hole."

Boeing was obligated to strengthen the floor and separate the control runs. Curiously enough, a United Airlines 747-100 at 22,000 feet over the Pacific lost a huge section of the fuselage, just behind Section 41 (clearly from Section 42) and the plane landed. This accident occurred in February 1989 on a flight from Los Angles to New Zealand. Fortunately the pilot managed to land the plane at Honolulu, but the accident did claim 16 lives. It could have been another major disaster like Pan Am 103. Why didn't Pan Am 103 land safely, if it had a similar hole in the side of the fuselage?

. . . And Failures

The story of Air India 182 brings up some important observations and serious questions. On Sunday, June 23, 1985, Air India Flight 182 was about 2 1/2 hours from landing at Heathrow Airport, at about 8:33 AM. At 7:13 AM, the plane vanished from the radar, even while the aircraft controllers were in contact with the pilots. There were 329 people on board, who perished as the plane sank 6,700 feet to the bottom of the Atlantic. Although much has been written about this accident, I recommend *Air Disasters. Dialogue from the Black Box* (Stanley Stewart).

According to Stewart, in July, 1985, the vessel Gardline Locator was on site with its robot mini-sub and recovered the black boxes. "On July 16, the CVR (Cockpit Voice Recorder) and the FDR (Flight Data Recorder) were opened in Bombay and their contents analyzed in the presence of international safety experts. At precisely 01:10.01hrs, the exact moment of the breakup, both recordings stopped abruptly. Flight 182's electrical power supplying vital components had been completely and instantly severed. — Whatever happened at 31,000 feet over the Atlantic was sudden and catastrophic indeed." Then, on the next page, ". . . with over 600 747's flying in the skies of the globe daily, could there be some kind of catastrophic structural failure? If such an event had occurred, other 747's throughout the world could be at serious risk."

The recovery work continued throughout August and September and wreckage was strewn over an area of five to six miles. The next point is extremely important: *"The forward aircraft section (Section 41) lay inverted at the beginning of the wreckage trail and was badly damaged"*! The investigators did not come to agreement as to an explanation of the sounds heard in the last seconds on the CVR, but many suggested that they were more like what one might expect if a cargo door had blown off than an explosion. The Canadian RCMP is convinced it was a bomb, and has spent tens of millions of dollars on its investigation.

The last chapter of *Air Disasters* says, "Before the concluding report could be published in 1986, however, the integrity of the 747 design was again cast into doubt with further revelations in January. In Tokyo, JAL engineers, still carrying out thorough checks in the aftermath of the disaster at Mt. Osutaka,* discovered cracks and broken ribs in the frames of the forward section (Section 41) of certain 747 fuselages.

To be economical, commercial planes are often airborne between 16 and 20 hours each day, 7 days a week. The number of take-offs and landings that occur each day is even more important than the hours aloft, as these are the times when the planes are under the greatest stress. These are known as "cycles."

The FAA issued an emergency directive requiring a worldwide inspection of the entire fleet. At the time, 610 747's were being operated by 69 carriers. All aircraft having completed 10,000 flights had to be examined within 50 landings. The FAA stressed the importance of the directive, the stating that "failure of adjacent frames could lead to rapid decompression and could

---

*On Monday, August 12, 1986, Japan Airlines Flight 8119 (a Boeing 747 with 528 passengers onboard) crashed when the rear pressure bulkhead collapsed. This problem is different from the one we are discussing.

possibly cause the loss of an airplane". Amid some alarm, more 747's with nose frame cracks were found by other airlines.

Then, in the next paragraph: "The revelations of forward frame cracking raised further questions as far as the Air India crash was concerned, and prompted suggestions that the VT-EFO's nose section, still lying on the Atlantic bottom, should be raised to dispel any doubts. The size of the structure, however, would make lifting very difficult. Engineers also stated that *the nose frame cracking leading to explosive decompression in an aircraft so young as the Air India was 'extremely improbable'. VT-EFO was only seven years old and had flown 23,634 hours in only 7,522 flights.*" That is an interesting comment, since on page 108, *Giant Jetliners* says: "The unusual bulb-shaped nose (Section 41) of the 747 created problems for Boeing, particularly in the early model. Some of these were not entirely unexpected. Boeing deliberately beefed up some parts of the nose section because it knew that nature prefers round-pressure vessels, not the 747-nose-shaped vessels with some flat surfaces. *Despite the strengthening, cracks were detected in the forward fuselage area known as Section 41. Some cracks were found on aircraft with as few as 6,500 cycles on the log book.*"

During my visit to the Boeing plant in Everrett, Washington, I verified that Section 41 is manufactured on one side of the gantry and then married with the fuselage on the other side. These planes are manufactured to such a high degree of accuracy that one section could be manufactured anywhere in the world and, when it comes to final assembly, the parts would fit together exactly. Section 41 is bolted and riveted to the main fuselage of the plane and the integrity of the connection relies on the quality of the materials used. If any parts are defective, the integrity of the plane is compromised; and that is the case with the early Boeing 747's.

What happens when the metal parts begin to fail? Of course, it would be a nightmare for the passengers and a catastrophe for the victims' relatives.

The problems with the Boeing 747 Section 41 were very well known to the industry, the FAA and the government, but not to the traveling public. In *Air Disasters*, the author points out: "In Tokyo, JAL . . . discovered cracks and broken ribs in the frames of the forward section of some 747 fuselages. The disclosure was another blow to the prestige of the 747, just as consternation over the 747 crashes in 1985 was beginning to subside".

The 747 was originally designed with the flight deck placed atop of the fuselage to permit nose-loading of cargo containers; this is the reason for the unique pear-shaped forward cross-section. The cracks were attributed to fatigue caused by cabin pressurization stresses on this rather singular structure.

The Pan Am Report

The Pan Am 103 Accident Report N0 2/90 (EW/C1094) now fits into the picture. The whole report was posted on the Internet at the website www.open.gov.uk/aaib/n739pa.htm. Under the general description, 1.6.2, it is noted that the working pressure inside the Boeing 747 would be 8.9 psi, which applies a considerable force on the frames and skin of the plane (like a balloon trying to form a circular shape). I quote, "in order to preserve the correct shape of the aircraft under pressurized loading, the straight portions of the fuselage frames in the region of the upper deck floor and above it [this is Section 41] were required to be much stiffer than the frame portions of the lower down in the aircraft." (The lower part is where the luggage is stored in containers like the one that allegedly contained the bomb that brought down Pan Am 103; the straight portion is caused by the fuselage being oval in Section 41). "These straight sections were therefore of much more substantial construction than most of the curved sections of the frames lower down and further back in the fuselage. There was considerable variation in the gauge of the fuselage skin at various locations in the forward fuselage of the aircraft."

It goes on, under 1.6.5 Maintenance Details. "The Pan Am 103, N739PA first flew in 1970 and spent its whole life with Pan American World Airways. At the time of the accident it had completed 72,464 hours flying and 16,497 flight cycles." (See the NTSB hearings for details as to what this means.) The plane had clearly undergone considerable maintenance and was in compliance with the airworthy directives at the time of the accident. Of course, the plane was still faulty — what it really needed was a new Section 41.

The Accident Report assumes the plane was brought down by a bomb and points to structural failures caused by the explosion.

Although I am not totally convinced that a bomb brought

down Pan Am 103, I am prepared to accept that it triggered the event and a structural weakness caused the ultimate demise. Much of the following discussion relates to what happened during the breakup that followed the explosion.

However, if one were to assume that Section 41 broke off without an explosion, exactly the same analysis would apply. I will not challenge the accepted theory at this time, but I will mention that in 1.12.3.3, "General damage features not directly associated with explosive forces", the report states "that a number of features appeared to be part of the general structure breakup, which followed from the explosive damage, rather than being a part of the explosive damage itself. This general breakup was complex and, to a certain extent, random." Big surprise! Later it states, "on the left side appeared to be the result of in-plane bending in a nose-up sense." Now, that is really interesting. Imagine what would happen if the lift on the nose, at 550 MPH, caused the nose to break off and upwards as I suspect it did, and as the accident report suggests.

What Happened to TWA 800?

As for TWA 800, the airplane would have been climbing to its transAtlantic cruising altitude of between 30,000 and 36,000 feet, depending on the prevailing Gulf Stream winds. The pilots had probably cleared the local traffic control and were settling in for the five or six-hour trip to Paris on what seemed to be a routine flight. At some point along the joints between Section 41 and Section 42 of the main fuselage, one or more of the bolts or bulkhead members was finally fatigued to the point that minute fractures and crystallization of the metal had already begun.

Fatigue cracks on the ribs of Section 41 and parts of the outer skin were now in an advanced stage and ready to fail at any moment, not unlike the United Airlines 747 over Honolulu in February 1989. Suddenly, the nose wheel door failed and tumbled to the sea, exposing the hole in the nose to the full force of the oncoming air. This in itself should not have caused a total decompression, as the wheel well is pressure-isolated from the main cabin. What did happen, I suspect, is that the force of the oncoming air, now hitting directly onto this open area instead of flowing around the smooth contours of the outer skin, applied irresistible forces to the already fatigued frames, ribs and skin.

I seriously doubt that either the pilot or the flight engineer knew what happened next. First, one member would fail — with a loud crack — and instantaneously this fracture would overload another already weakened member; in a fraction of a second, the entire Section 41 would be separated from the main fuselage.

You will recall that the nose section of the plane, at 550 MPH, is helping to lift the plane; thus, when the break finally came, this upward force caused Section 41 to burst upward with an enormous force. The entire cabin would suffer an immediate decompression, which would be felt by all as the ears burst and the lungs collapsed. The main part of the plane — minus Section 41 — would be tail-heavy and would sink backwards to some extent, even though it was traveling forwards at 550 MPH. This will explain why Section 41 did not hit the tailplane as it careened upwards and over the fuselage. Section 41 would then tumble down toward the ocean, with its pilots still strapped in their seats, having been rendered unconscious by the excessive G-forces and other traumas. The remainder of the plane would still be trying to fly, with its engines still running at high speed. The hapless plane might climb, at first, then dive like a porpoise, desperately trying to fly as it was originally designed to do — until the wind tore the plane to pieces.

The huge front of the main fuselage would now be experiencing a 550 MPH wind slamming directly into the cabin area; such force would split the cabin in two. At the same instant, the violent gyrations would cause parts of one or both of the wings to shear off, releasing thousands of pounds of fuel into the air, where it would immediately catch fire. Contrary to the scenarios offered by the FBI, I seriously doubt that the small amount of fuel in the central tank would be seen streaming from the fuselage as it plunged towards the earth. More likely, it was the wing tanks that the residents of Long Island saw burning.

The forces were too great to have allowed the plane to stay intact for many seconds after the Section 41 separation. This would explain why the parts of the plane were found over such a wide area; some 800 square miles in the case of Pan Am 103, and 225 square miles of ocean in the case of TWA. When the plane finally broke up, the pieces were still traveling forward at very high speed, albeit less than 550 MPH. The inertia of the parts would carry them in many directions and over long distances, depending in each piece's weight and wind resistance.

The NTSB analysis forcefully explains that the explosion in

the central fuel tank caused the front of the tank to hit the bulkhead, where damage could be clearly seen. However, they ignored the alternative explanation whereby the bulkhead had been forced backwards by the power of the oncoming air and hit the central fuel tank. This is a much more likely explanation.

According to the public announcements by the NTSB, the FBI and indeed the CIA, the central fuel tank exploded, causing the front one-third of the plane to separate from the area just ahead of the wings. This would be Sections 41, still joined to Section 42 as it plunged to the ocean. In the animation of the crash produced by the CIA, the narrator clearly states that the front third of the plane plunges to the ocean, and the animation is visually very clear too. However, if you read the accident report the story is quite different!

In article 2.1.1, "Section 41 and Forward End of Section 42", the report states: "The forward section of the fuselage from STA 90 to approximately STA 840 comprised of Section 41 and part of Section 42, was found in the Yellow debris field." What appears to have happened was the total failure of Section 41, which pulled part of Section 42 with it.

In the case of Pan Am 103 and Air India 183, Section 41 broke free without dragging a part of Section 42 along. In the case of TWA, it would not be unreasonable to accept that part of Section 42 went with Section 41.

The above discussion explains what happened more plausibly than the three theories that were widely disseminated. Why do the investigators have so much trouble figuring out the real cause? I sincerely hope the people in charge were not aware of the "Boeing Report" and were acting in good faith, but I have my doubts. Some, like FBI Assistant Director James Kallstrom, who headed up the FBI investigation, appeared to be very dedicated and honest in their analyses. As for others involved in the investigation, I have very grave doubts.

*

TWA, the government, and Boeing must have been on red alert. The damages could be enormous, in the billions of dollars. How did the various groups that should have been aware of the real situation with these early Boeing 747's respond to the situation?

1. Transworld Airlines: Immediately after the TWA 800 crash, Mr. Mark Abels, Vice President of TWA, was denying any responsibility whatsoever and was eagerly suggesting terrorism, missiles or other possibilities, which might have absolved the company from any liability. In one TV documentary, Mr. Abels appeared four times answering questions about the crash and clearly scoffing at any inference that TWA might know what caused the accident.

I actually agree with Mr. Abels' suggestion that the fuel tank was not the problem. However, he fails to mention that this old TWA Boeing 747-100 was scheduled for a major Section 41 repair program later that year. The parts had already been ordered in March 1996 — four months before the crash.

Then TWA came up with an even less credible position, suggesting that the crash was over international waters, thus covered by a very old law of the sea, limiting its liability to a very nominal amount. The victims' relatives cried foul and the US

Congress got involved, changing the law retroactively.

2. The Federal Aviation Authority: There is little question that the FAA knew what was going on, since it had already directed the required repairs and approved the timing of the work. These details are quite clear in the "Boeing Report" and documented in many other books on aircraft crashes, which specifically detail the problems with Section 41. (See Bibliography).

3. The Boeing Corporation: The Boeing Corporation is an amazing organization, clearly staffed with some of the brightest minds in the aircraft manufacturing business. The company has a very long history of major achievements, which spans back for over sixty years.

However, Boeing is known to be tight-lipped, and not only concerning the Boeing 747 problems. It has been sued by its investors for not being forthcoming with its financial statements. Boeing had no incentive to volunteer any worrisome tidbits on this occasion.

4. The Clinton Administration: *The American Spectator*, no friend of the Clinton Administration, published an article by John B. Roberts II in July 1997, discussing Al Gore's involvement in the TWA 800 accident.

> When he was putting together his White House Commission on Aviation Safety and Security late last July (1996), after the explosion of TWA 800, Bill Clinton took the trouble to call Victoria Cummock personally and ask her to join. Cummock, who had lost her husband over Lockerbie on Pan Am 103 eight years before, seemed a solid, logical choice. Since her hus-

band's death she had devoted so much time and effort to improving airline safety that *Newsday* labeled her "the airlines' most tenacious foe." Aboard Air Force One — en route to a New York appearance to share the grief of the families of TWA 800's victims — the President convinced Cummock that he sincerely wanted to develop stringent new counter-terrorism measures for America's airlines. Vice President Gore, he added, would head up the new Commission. She agreed to join. Since that time, Victoria Cummock has filed a suit in Federal Court against Gore and the Department of Transport, charging that the Vice President pressured her to abandon her call for counter-terrorism measures and refused to publish a 42-page dissent she had filed. "Al Gore, she believes, sold her out."

So how much does the Administration know about these crashes?

5. The United States Senate and Congress: In another *American Spectator* article, August 1999, by John B Roberts II, the following quotation might shed some light on the position of some members of Congress.

> It was billed as an investigation of the investigators. On May 10, 1999 Sen. Charles Grassley (R-Iowa) held a one-day hearing with witnesses offering damaging testimony about the Federal Bureau of Investigation's role in the TWA 800 probe. Grassley's opening remarks were particularly critical of former FBI Assistant Director James Kallstrom for failing to uncover the cause of the explosion that killed the jumbo jet's 230 passengers and crew on July 17, 1996.

Kallstrom rebuts the charges as "a bald-faced lie." Later in the article, "Grassley's headline-grabbing probe scapegoated Kallstrom as though he had operated on his own authority, depicting him as a rogue cop running out of control. At best, Grassley's portrayal of Kallstrom is a caricature. At worst, it is character assassination."

I believe Kallstrom took a very personal interest in the investigation and is to be highly commended for his effort.

On pages 42 and 43, this same article states,

> Neither Gassley nor Duncan pressed Hall (James Hall, chairman of the NTSB) to answer still unresolved questions on Hall's and the President's roles in raising more than $500,000 in soft money contributions from the airline industry for the 1996 Clinton-Gore re-election effort-a time when the White House Commission on Aviation Safety and Security was considering security measures which could have cost the industry $1 billion.
>
> ... NTSB Chairman Hall's high-level briefing for reporters also undermined the FBI. Hall and other NTSB senior staff ridiculed competing theories of the case, misleading the press and leading the public to believe that only the accidental-explosion theory, for which there isn't any conclusive evidence, could explain TWA 800's destruction."
>
> ... What the CIA did not explain in November was that its video was altered after consultation with the NTSB [on November 18, 1997, the CIA had produced an animated video simulating TWA 800's final flight]. In a letter from the CIA Director

George Tenet to Rep. James Traficant (D-Ohio) dated January 13, 1998, *Tenet acknowledges that more than forty changes were made to the video animation at the NTSB' suggestion. After the changes were made, Tenet says the CIA showed to the video to "NTSB managers" who approved its release to the general public.*

On page 45 the story continues. "Kallstrom is disappointed that FBI director Freeh didn't speak out against the Grassley hearing, which Kallstrom calls a "Kangaroo Court." He is not alone. Two weeks after Grassley's hearing, on their own initiative, some 400 FBI agents and professional support staff from the New York office sent Grassley a letter protesting the hearing as "one-sided, incomplete and distorted." Finally, this well-researched article concludes, "Hall wants the case sewn up before the campaign season (1999) gets underway".

The National Transportation Safety Board: On December 5, 1997 the NTSB held a week-long hearing in Baltimore, Maryland on the situation of TWA 800. This was a very unusual hearing as the NTSB rarely makes such public display of its work. I will say up front that the whole effort appeared to be a white washing sham. It was broadcast by C-SPAN. Was Boeing going to tell the world about the problems with the Boeing 747 Section 41 fatigue problems?

Instead, it seemed that Jim Hall, Chairman of the NTSB, had decided, for whatever reason, that an explosion in the TWA 800 central fuel tank had caused the crash. He controlled the meeting, and his view prevailed, adding to the emphasis on the fuel explosion theory.

The big problem with the central fuel tank explosion theory was to determine what had sparked the explosion. The meeting was clearly focused on the potential problems with the electrical wires. There had been many incidences with wires becoming exposed in the older planes due to wear and tear. Hall made sure that a lot of time was spent on wires and the probes that enter the fuel tank pumps and instruments, which measure the quality of fuel in the tank.

For those unfamiliar with such systems the whole procedure would be impressive. The observers' attention was directed to the fact that the wires connected to the fuel measuring probes were in the same bundle as other wires with much higher voltages. To be specific, the fuel probe voltage was designed for five volts while other wires in the bundle were at 400 volts. If these wires came in contact through wear, then there would have been a spark — a big one.

However, this could not have occurred inside the probe or even near it. If a wire carrying 400 volts touched a wire with five volts, the five-volt wire would immediately melt and cut the power to the probe. The pilot would have noticed the electrical power cut immediately — the fuel gauge would have read zero! There is no circumstance in which a five-volt wire could carry 400 volts to the fuel probe and cause a spark in the tank.

The only experts on the subject were Mr. Ivor Thomas, fuel systems expert, and Mr. Jerome Hulm, expert on the Boeing 747 electrical systems. Both men were from Boeing. They gave very long and detailed explanation of the systems and appeared to be answering the questions directly, honestly but carefully. Their testimony was technically honest — but unconvincing.

It must have been well into the fourth day that Robert Vannoy, 747 Fleet Support Chief, brought forth his startling presentation, which lasted a little over 30 minutes.

Vannoy, a polished, articulate individual with small eyes and a controlled expression, was clearly in a hot seat, but only his occasional downward gaze showed that he was less than happy with what he was saying. The body language was not unlike that of Mark Abels when he was uncomfortably explaining the TWA position.

Vannoy's subject was the evaluation of "Aging Aircraft." He started out by telling the panel that he had a ten-page presentation. I knew then that Boeing was not about to show the extent of the problems with Section 41, at least not graphically.

The first Boeing 747's were brought into service in the early 1970's and by 1980, some planes had reached their original design limits and some planes had shown signs of metal fatigue. Boeing became concerned.

The next section was entitled "Aging Watch" and Vannoy told the panel that Boeing decided that there was no real problem with the older planes, as long as they were maintained. Originally the Boeing 747's were designed to perform for twenty years, or 20,000 flight cycles (one flight cycle is one takeoff and one landing) or 60,000 hours of flying time. At the time of this hearing, 380 Boeing 747 had exceeded 60,000 hours in the air and 240 exceeded 20 years in service and 20,000 cycles.

The question came up: what was the "age" of TWA 800 when it crashed? Vannoy answered, "90,000 hours, 25 years old and 19,000 cycles."

Clearly, the TWA plane had spent most of its time in the air, as the flight cycles were relatively low compared to the air time, compared to other aging Boeing 747's. It must have been flying long journeys most of its life. The most stress on any plane has to be on takeoff, when it is fully loaded with cargo, passengers and fuel, but the time of this stress is very limited. The sustained stresses on a commercial plane occur when the plane is flying at altitude with the cabin pressurized.

Vannoy went on to say that about 700 Boeing 100's, 200's and 300's were of the older design, and 620 of them were still flying. 30 planes had reached limit of their economic life and had been scrapped; he did not say what had happened to the remaining 50 planes. In fact, the total number of Boeing 747's was not 700, but 686. After the 686th plane had been built, a major design change was introduced.

Vannoy put up another chart called, "Supplemental Structure Inspections." Boeing 747 planes were beginning to experience fatigue problems and in 1983 Boeing introduced a program consisting of 20 reports called "Airworthiness Directives." These directives required an immediate inspection for metal fatigue.

120 planes were checked, and fatigue problems were quite evident.

Jim Hall asked Vannoy to explain metal fatigue cracking, to make sure that all aspects of the hearing had been covered. Vannoy was visibly uncomfortable. He was clearly "off script". He muttered, "1/10-inch cracks, skin, lugs, small cracks and we find them in the early stages of development". Hall was satisfied and asked no more probing questions.

Then Vannoy pulled up the next chart, "Redundant Structures." He said that the teams checked 100 places for metal fatigue, trying to find the first signs of any cracking. They found twenty areas that required inspection and sent that information to the FAA and the air carriers. Then Bob Swain, of NTSB Aircraft Systems Investigations, asked Vannoy how Boeing found out about the cracking. Vannoy answered that the air carriers reported problems to Boeing and Boeing told the FAA! That's a nice, cozy relationship.

He was then asked, "How do the carriers know where to look?" Vannoy coolly told the panel that the carriers are given a "Zone Diagram". Why didn't Vannoy display the Zone Diagram? I have a copy from Curtin, the FAA has copies, the air carriers have copies — why can't the public and the victims' family members who were present at the Hearing see these diagrams?

Even more suspicious, Dr. Bernard Loeb, a frequent guest on the documentaries and other TV programs, did not jump in with more probing questions. He was easily satisfied, as was Jim Hall, that Boeing was doing everything it could to address the metal fatigue problems. Vannoy did put the FAA in the hot seat when he made it clear that Boeing was telling the FAA all about the problems and it was up to the FAA to inform the air carriers. The FAA was silent.

Vannoy moved on, bringing up the next slide, "Fleet Survey Program Findings." This program, in operation from 1986, looked at metal fatigue problems with Section 41 and Section 42 (with emphasis on Section 41). The main structures were experiencing very few problems compared to Section 41. Vannoy said the Section 41 problems were the best known, and that was in agreement with third party studies.

Hall jumped in again. "Please elaborate. Can we see a diagram of Section 41?" Here again, Boeing could have presented a detailed schematic of Section 41, but failed to do so. What was shown was a drawing of the entire plane, showing the wiring problems and someone pointed to the nose. . . Hall was happy with the explanation.

Vannoy was trying to be up-front with his remarks but a picture would have been worth a thousand words. *Vannoy then told the panel that in 1986 Boeing knew they had a problem, an urgent problem, and that all planes with more than 15,000 flight hours needed an immediate inspection. Boeing was alarmed and made some serious design changes to models after number 686 had been delivered. All the older planes needed to be retrofitted.*

Then came the big one. Hall asked if TWA 800 had been retrofitted. Vannoy said that it had not — but was due for servicing. It had just over 18,000 cycles and was due to be fitted at 19,000 cycles. Cracking on the TWA 800 had been seen, but the TWA management had not agreed to apply the retrofit yet.

The retrofit parts were to be ordered in March 1996, some four months before the plane crashed. The fact that TWA chose the number of cycles as the only criteria on which the retrofit was to be undertaken was a deliberate money-saving decision by a company in serious financial difficulties. The other criterion, whereby the plane would have been retrofitted after 60,000 flying hours, was ignored; TWA 800 had 90,000 flying hours. Why didn't Jim Hall blow his stack?

Then someone jumped in and said that they had looked for fatigue cracks on the TWA 800 wreckage and found none. What? The plane was in a million pieces, and the NTSB was looking for old fatigue cracks. How could the NTSB have seen the difference between old and new cracks? Why did this individual suddenly jump into the conversation to make such a forceful comment? Was he concerned that the Hearings were getting a little to close to the truth? Hall brought the story back to the wiring problems but Vannoy had not finished.

With his next slide, "Alert Service Bulletin", Vannoy told the audience that Boeing acquired a used Boeing 747 with 20,000 hours flight time over two and one half years, and simulated an additional 20,000 hours. *This result triggered the Alert Service Bulletin, which is an alert with a much higher priority, signifying safety considerations.* Air carriers were all given the information. It stated that at 22,000 cycles, major retrofitting was required.

Nobody asked any more questions, so Vannoy put up another slide: "Diagram of Service on Alert." Hall tried to move back to the wiring problems, but Vannoy still had not finished. He stated that 40 planes had undergone the retrofit and the conversation moved to the Aloha 737 problem over Hawaii. Corrosion was the subject and Vannoy was clearly on the home stretch. He had made it through his presentation with hardly a murmur about the real gravity of the Section 41 metal fatigue problems.

Legally, Boeing told the whole story (in very brief terms),

considering that the fatigue problems had been identified in 1980 and a serious program had been in operation since 1986. It should be pointed out very clearly that Boeing has absolutely no facilities for repairing any of its planes. All repairs are conducted by authorized third party maintenance companies — Boeing simply provides parts and advice.

The swirl of obfuscation around these issues continues to grow thicker. In September 26, 1999, *The Sunday Times* run an article about British Airways trading in some 50 Boeing 757 planes for new Boeing 717 planes. There are two points of interest in the article. First, it states that Boeing is in the best position to refurbish the older planes — which, in fact, it cannot do. Secondly, it states that BA is switching from Boeing 747's to Boeing 777's. You will recall that BA did not like the oval shape of the 747 and would not buy the planes until changes had been made.

Has the truth about TWA 800 ever been told?

Chapter 9

LATE BREAKING NEWS

Cromer crabs are not only a local delicacy, but are welcomed all over the United Kingdom. Each morning I watched the sun rise over the Cromer pier and observed the crab fishermen hauling up the catch of the day, while I sat at my laptop computer hauling in information from all over the globe using the Internet.

On November 5, 1996 I was in the little commemorative building in the churchyard at Tundergarth, signing the guest book to pay my respects to those who were lost in the crash of Pan Am 103. Now, as I continued to write up my assessment of events, the crab men were busy in the early rays of light, still at their task as they had been for over 1,000 days since my first visit to Lockerbie and as their forefathers had been for 1,000 years before.

Three years had passed since my Lockerbie story began, and that story as well as the TWA 800 and Air India 182 crashes were still active. Among the most tantalizing events still to unfold is the trial of the two accused Libyans, was set to begin in May 2000 in Zeist, Holland; in its aftermath, certainly, more intriguing bits of information will continue to circulate, although they may confuse the picture as much as they illuminate it.

This book is intended to be balanced, impartial, and based on proven facts, with special value added by the Boeing Report that I obtained from Curtin. On the basis of my extensive investigations, I must point out certain discrepancies and follow-up information that casts more, or less, credit on the words of some of the people who became well-known in the Lockerbie story.

The documentary *The Maltese Double Cross* is a very powerful piece of investigative reporting and I have confirmed much of the content through corroborating evidence. However, although the words from Mr. Oswald Le Winter might well be factual, his subsequent diatribes landed him in jail. I read on the Internet that "Self-styled CIA agent Oswald Le Winter [had been] jailed for two and one half years by an Austrian court for trying to sell fake documents to Harrods owner Mohamed al Fayed, purporting to show that Princess Diana and al Fayed's son Dodi were murdered by British Intelligence agents." If Le Winter tried to cheat al Fayed out of $15 million, then I should point out that his testimony in this case is highly suspect.

In September and October 1999, renewed attention accrued to the legacy of President Ronald Reagan. *Dutch: A Memoir of Ronald Reagan*, by Edmund Morris, was featured in *The Sunday Times* October 3, 1999 and in other papers. It was Reagan who had nominally initiated the big quarrel with Libya by bombing

Tripoli in 1986, ostensibly in retaliation for what was said to be Libya's role in the bombing of the German discotheque *La Belle*. Whether Libya was or was not involved in that attack has never become completely clear. The French and Spanish were against the U.S. counterattack.

Richard Cohen wrote a piece for the October 6, 1999 *International Herald Tribune* wherein he said, "Here is the Reagan we all knew — boring, out of touch, intellectually lazy, blessed with good luck, good looks, a sunny disposition and 'ambition enough to crack rocks.'"

If Reagan was "intellectually challenged", then who was behind the vast job of running the presidency of the United States of America? Who was really behind the dangerous games of KAL 007 (rumored to have been part of a conscious and systematic use of civilian aircraft for military espionage work), that almost triggered World War III? Who was behind the Iran-Contra affair and other troubling events that took place on his watch? The Libyan trial, it is hoped, will shed some light on these questions.

Ever since Gadhafi committed the two accused Libyans to stand trial in The Hague, he has been on a crusade to improve both his own international image and that of his country. He has entertained many African leaders, and hosted a specially convened summit meeting of the Organization of African Unity during which he denounced the other Arab countries of failing to support him during the U.S./UN sanctions. Now Gadhafi has shifted his focus to the creation of a United States of Africa. The mercurial Gadhafi has sought world attention with many attempts at unions with Sudan, Egypt, Syria and others — all these efforts came to naught, squandering Libyan resources. The war with Chad was an unmitigated disaster and his financing and smuggling of arms to the IRA probably encouraged Margaret Thatcher to support Reagan in bombing Libya, in 1986, even though the Boston Irish were openly doing the same thing.

UTA's DC-10

When Gadhafi overthrew King Idris on September 1, 1969, the first people he forced out of the country were the Palestinians. However, he later became an ardent supporter of an independent Palestinian state. There is little doubt that Arabs were trained in guerrilla tactics in Libya — but then, many countries train troops, sometimes in their own countries and sometimes in foreign lands.

Despite Libya's vast oil revenues, Gadhafi knew he could not win a battle with a major power; thus, his only option would be guerrilla warfare. Whether Gadhafi initiated the bombing of the French DC-10 (owned by UTA) in his guerrilla war against France — in response to its support of Chad against Libya's interests — has been brought into question.

The UTA DC-10 exploded over Niger on September 19, 1989 while on a flight from Brazzaville, Congo, to Paris. All 170 people perished. A French court tried and convicted six Libyans in absentia, including Gadhafi's brother-in-law, Abdallah Senousi (reportedly the head of Libya's security services). Gadhafi has publicly stated that Libya would be willing to pay compensation to the families of the victims of any airline bomb attack, if a competent court ruled against any Libyan national as the perpetrator. When the UTA verdict came down against the Libyans, he paid France 200 million francs. This payment was immediately construed as an admission of guilt and the French demanded Gadhafi's head. Interestingly enough, the French prosecution said that these six Libyans were also involved with the Lockerbie bombing. If it turns out that Libya was not involved with the Lockerbie case, where will the UTA case stand?

A documentary called *Covert Action* was aired on October 3, 1999 on BBC2; the whole hour is devoted to President Reagan's obsession with anti-communism (a noble cause in itself, but manifested in a rather unsavory manner). In the same period when Reagan bombed Tripoli, he and CIA chief Bill Casey attempted to embarrass the Russians, who were at war in Afghanistan. Reagan wanted the Russians to lose, or at least to get bogged down in a Vietnam-style conflict that would consume much of the communists' resources and efforts. On the other hand, Reagan knew that direct involvement would receive very negative public reaction, so he called on Maggie Thatcher. Thatcher and Reagan had played the good guy/bad guy routine across the world, the Falklands war being a case in point.

The British began shipping arms, old arms, to the Mujaheddin, the Afghan freedom fighters, via a covert operation in Pakistan. The Afghans were clearly outgunned by the Russians, who attacked them mercilessly with gunship helicopters. To help the Afghans, Reagan pushed Thatcher to send weapons that could shoot down the helicopters; she obliged by sending the useless "Blowpipe" hand-operated ground to air missiles. The Afghans were told that the missile launchers were effective and when they failed, the Russians simply blew away the Afghan forces. Finally, Casey and Reagan decided to send U.S. "Stinger" missiles; this caused real havoc with the Russian offensive. Finally, the Russians were economically and physically wiped out.

Suddenly, this was no covert action, the U.S. was clearly involved. Was Reagan on the brink of provoking a massive reaction against the USA, as he had almost done with the KAL 007 incident? In any event, handing over hundreds of "Stinger" missiles to anti-Western Arabs, including the armies of Osama bin Laden, clearly endangers the West. If that sort of adventure was typical of Reagan, then the suspicions about KAL 007 become more plausible.

Fly, Britannia

On September 16, 1999, the British paper, *The Sun*, published a photograph of the Britannia plane that crashed in the Costa Brava while trying to land during a thunderstorm. The picture shows very clearly how the plane broke — cleanly, just in front of the wheels and just behind. The front break looks very similar to the break that occurs when Section 41 on the Boeing 747 severs from the Section 42.

A Bit of A Scrap — Jim and Jane Swire

The Swires lost their daughter, Flora, on Pan Am 103; Jim was elected to represent the UK and Scottish victims' families.

When I went to visit them, I found Jim very engaging and we quickly developed a warm relationship.

The first meeting was mostly to get to know each other; Jim was careful to spell out his particular situation with respect to the upcoming trial in The Hague. He was obliged to reveal very little, and had decided to report anything of significance to the Lord Advocate, Lord Hardie. Jim and I were able to discuss anything that was in the public domain, however, and so we did.

During our first meeting, Jim told me that David Ben-Aryeah developed a very complete history on the Pan Am 103 disaster (he had featured very eloquently in *The Maltese Double Cross*). I contacted Mr. Ben-Aryeah, via Internet, to ask if he had enough clout to obtain two scraps of aluminum from the Pan Am 103 cockpit, which was rotting in Tattershall. If the cockpit, I theorized, was so unimportant to the case, why couldn't he see it and get a small piece for analysis? I had tried to have an investigative reporter get these small samples and he failed; but one can never know how persistently he had tried.

If Mr. Ben-Aryeah were so interested in the case that he had spent years on the subject without pay, then one would have thought he would have jumped at the opportunity. I was shocked to get an angry e-mail effectively telling me to get lost. Maybe he was writing a book, and did not want any further competition? Jim did tell me that the cockpit would probably go on the market for scrap after the trial was over, indicating that I might be more successful later. I dropped the subject. Let the next interested writer pursue that angle.

Bollier, Or Odd Timing

On October 16, 1999, I set out to visit with Mr. Bollier, of MEBO and the Bombeat Radio.

I flew EasyJet from Luton, UK, to Zurich. I've been in cleaner and better-organized junkyards than the Luton airport, and the endless warnings of health risks associated with smoking had clearly been lost on Luton passengers. EasyJet flights are advertised as cheap, but that only applies if you book years in advance, stay over Saturday and the following Tuesday night, and don't expect coffee or a decent seat. If you're on business and need to travel immediately, the agent will kindly explain that they have you over a barrel; you can always take the train. Next time, I think I will.

Cheap British flights attract people who never go to a dentist or hairdresser and who rarely bathe. EasyJet has a system of random seating, which means that the British, who will usually line up in an orderly fashion for no reason whatsoever, will trample over your kids in a desperate effort to board. Once on board, you are captive; the staff will make use of the flight time to try to sell you various items. And you had better have the currency-of-the-day, because "foreign" money is only accepted under very adverse conditions.

I was lucky. I got an aisle seat, up front, so I could disembark without being crushed by the crowd. Of course, I had to accept the permanent indentation of the aluminum bar, upon which I had been sitting, into my upper thighs. The folks across the aisle must have been on their first flight — even though the flight was only 90 minutes, they promptly unpacked a huge lunch, which they ate with manners that were barely less appealing than the contents of the meal. It was good to be going back in Switzerland, but at what a price!

I checked into the Intercontinental Hotel on Badnerstrasse, took a long bath to ensure the lice were all drowned, and called Ed. He agreed to come over at 18:00 hours and chat; like all Swiss, accurate as their clocks and watches, he came on time. I quickly forgot my travel ordeals and sat mesmerized at the hotel bar as Ed recounted his experience with the Scottish prosecutors the week of September 13, 1999.

Ed Bollier and his partner, E. Meister, were in the spotlight due to their association with the MST timer, which allegedly triggered the bomb that brought down Pan Am 103. Bollier had, by this time, spent the better part of eight years searching for the truth behind the crash and whether or not his equipment was involved.

Unlike Dr. Swire, Bollier was under no pressure or restraints to minimize his dissemination of information. He had no illusion about the role he had been forced to play. The media had slandered him, he had lost his business and his house, the banks cancelled his business line of credit, and the death threats had been unnerving. He made the conscious decision to make sure that everything he knew was also known to the Swiss police, the Scottish police, and anyone else who showed a professional understanding of the situation (myself included). Much of the information was also posted on the Internet — a very wise move, in my opinion.

Bollier and his attorney had worked out a protocol with the Scottish prosecutor's office such that he was entitled to review all data and inspect the fragments of the MST Timer and the Toshiba 453 Bombeat radio. He was also entitled to discuss the situation with the defense team as well as with the prosecutors. This protocol unnerved the prosecutors, as was clearly demonstrated in a letter from Ms. Mirian Watson, Principal Prosecutor, Fiscal Deputy, to Mr. Bollier, dated September 8, 1999. "Precognition is a private interview between you and the prosecutor, and I assure you that this is my only intention in inviting you to Scotland."

When Bollier arrived at the prosecutor's office in Edinburgh, the situation was very tense. Bollier's attorney had repeatedly demanded to see the original MST timer fragment and had been repeatedly denied.

On Tuesday, September 14, 1999 at 4:15 PM, four policemen came into the room carrying a sealed phial, in which the fragment lay, inside a plastic bag. Bollier could not see the small fragment clearly, and asked that it be removed from the phial. The prosecutor refused. Bollier exploded and informed the prosecutor that he would write a new report, clearly stating that he had been refused the right to examine the fragment.

It was the prosecutor's turn to get upset; she left the room. Suddenly, it was agreed that the phial could be opened in the presence of the police, and Bollier was allowed to look at it using a powerful magnifying glass and specially adapted spectacles. For complete clarity, I quote from Bollier's sworn written testimony dated September 16, 1999:

"I established that the fragment was colored green, (lacquered) on one side, and identified it was manufactured by 'Thuring' Switzerland. This type of PC board was used in at least five MST13 Timers supplied to the Libyan Army.

When asked what I could see, by Ms Watson, Prosecutor Fiscal Deputy, in the presence of Mr. D Harvie of the Criminal Investigation Team Crown Office, I said that the solder points where a relay would normally be connected to the board showed that there had never been any soldering there (i.e. [there was] no structure). The edge of the solder point on the fragment that I was shown had no distance between it and the edge of the circuit board. Thuring PC boards normally have a one mm gap between the edge of the solder and the edge of the PCB. Seeing this, I said, 'This fragment cannot be from a working timer.'

Ms Watson then asked me why not and I said, "There is no structure on the solder point to indicate that a relay had been soldered to this point." For me, this fragment is 'fabricated.' By this I mean that it came from a PCB that never had any electronic component parts inserted into the board. This was heard by all those present during the examination.

I now requested that this fragment be examined by five American experts and one expert from RACAL in the field of electronics and physics to confirm that a relay was never soldered onto that fragment.

On 15th September 1999, during the course of the morning, an officer in a white knee-length coat appeared and showed me — in the presence of Mr Harvie and Ms Watson as well as the Interpreter — one missing bit from the fragment (the sawn-out piece). This piece was light brown, beige color, from the first series of 3 prototypes. At the same time he asked me to look at the green piece. This was no longer the one I been shown on 14th September, 1999. This fragment had two solder points and there was still no evidence that a relay had been soldered to it.

To follow this exactly, you need to look at the picture we obtained from the video taken when Mr. Thurman (at the FBI laboratory) was explaining how he discovered the evidence that led to the two Libyans being indicted. At first, Bollier thought that the fragment was from a timer he had sold to the Libyan Army. Upon further examination, he found that the PC circuit board he was looking at had never been made into a complete timer at all. Bollier and I went to his office, where he showed the picture of the fragment analyzed by Thurman on June 15, 1990. Bollier then demonstrated how the timers are made and how a reply is inserted into the board, and soldered. He explained that the relay has terminals, or pins, protruding from its base. Once these pins are inserted into the board, they are bent over to ensure a positive contact with the circuit and to avoid a 'cold' soldering. The fragment from the Pan Am 103 showed no evidence of this soldering. Bollier said he was sure that the fragment he was inspecting was from the film stolen from his office (which he had reported to the Swiss police long before the bombing). Then Bollier pointed out a very important aspect of the fragments.

One can see that the fragment has been cut. The larger piece was green and no burning was evident; the smaller piece was brown and burnt on one edge. These two pieces were not from the same board, even though they had been presented as the entire piece from the timer used in the bombing.

Fortunately, Bollier is a technical expert and a very careful investigator. I should point out that this fragment is only about 3 mm in size; they say it was found in the debris field around Lockerbie: some 845 square miles.

It gets worse. When Bollier saw this evidence, in Scotland, he was determined to make sure it was witnessed. Bollier asked one of the policemen to come and witness the two parts of the fragment, but he was very nervous and begged out — claiming he was working with the prosecution. Ms. Watson was getting very upset over these demands, but finally they agreed that a policewoman could inspect the fragment parts and mentally note the colors, but not say what she saw. Bollier was asked to leave the room while the witness wrote down what she saw.

Once all of these witnesses had completed their work, the fragments were labeled and Bollier signed each one, after having verified that they were the ones shown in the police photograph identified as PT/35(b) and DP31 on page 336 of the investigation files.

Bollier made several other very interesting observations on this trip to Edinburgh.

1. Mr. Ulrich Lumpert was the technician who had made the first three prototype MST 13 timers. The first two prototypes had been left with the Stasi in East Germany and the third,

according to Mr. Lumpert, was damaged and thrown away. However, it has been discovered that Mr. Lumpert made three trips to the USA while the timers were in the prototype stage, and as of October 1999, this was being investigated in detail.

2. It has been confirmed that Mr. Charles Buyers would testify in The Hague that he knows exactly where in Florida the MST 13 circuit board was manufactured.

3. Mr. Bollier has seen only a photograph of the fragment from the Toshiba radio allegedly used in the bombing of Pan Am 103. However, nobody at the Edinburgh meeting appeared to know where the fragment was. As I explained earlier, the Toshiba Bombeat 453 radio allegedly used in the bombing was similar to those found in Neuss, Germany when the German police arrested Dalkamoni and Kreesat. Yet Bollier discovered that the fragment could only have come from a Toshiba 8016 radio/cassette player, which was not called Bombeat. The photograph of the radio fragment clearly showed a part of the circuit from a Toshiba 8016 model.

4. Bollier then told me that two members of the Jaffa family had been on Pan Am 103 and only one of their bodies was found. The other body had been removed by the extremely active FBI agents within a few hours of the crash.

We will call Jaffa Number One, the body that was found near Mr. Cannon's body, near the cockpit at Tundergarth. Jaffa Number Two was a student who had visas for Lebanon, Germany, Sweden and the USA and had travelled to London via Cyprus, connecting to Pan Am 103 in London. Evidently, Jaffa

Number Two was seated at the rear of the plane and was found by a police doctor, who was tagging the bodies. Later, most of his official tags had been removed, and when the doctor prepared to sign an affidavit to this effect he was fired from the police force.

A curious development had taken place while the Pan Am flight was in the loading stage. Evidently Jaffa Number Two had been taken on board, circumventing the normal channels. At the same time a Mr. Whitaker, whose baggage had been loaded, missed the flight (he'd had one nip too many). How convenient! When the stewardess counted the passengers, the tally would have matched the number of people expected on board. In October 1999, a major investigation of the whereabouts of Jaffa Number Two was under way. Was he, too, working for Oliver North and shipping heroin to the USA under *'controlled deliveries?'*

On November 1999, I called Lord Weatherill — Lord Weatherill had been speaker in the House of Commons for years and had been made a Lord for these services to the British government and who, on previous occasions, had some pertinent words about the consequences of the public not participating in the public affairs — and he agreed to see me at the House of Lords. I had a delightful discussion, which culminated with an invitation to a very historic event whereby The House of Lords would be effectively self voting out power after 800-years of continuous service, on November 11, 1999 at the House of Lords. I was actually on the floor of the House of Lords for the two minutes' silence as the members remembered those lost in the two World Wars. This was followed by a lunch with Lord Weatherill. That afternoon I was in the gallery observing this historic event and, unbeknown to me at the time, Lord Weatherill was writing the words I wanted on the back of the document referring to these events.

On December 22, 1999, a Korean Boeing 747-200 crashed on takeoff from the Stanstead airport in the UK. I gathered as much information as possible. The downed Korean Boeing 747 had been the 448th plane built by Boeing and was licensed as HL 7451. This plane too was scheduled for structural modifications according to the "Boeing Report." No doubt this case will take on a greater significance and will require more intense investigation.

On March 14, 2000, I was invited to review a project in Ireland, which was being organized by an Irish and American company to burn household trash and generate electricity with the power. I was very familiar with the project from my experiences with Dardas and North American Technologies. We decided to follow up with the Irish entrepreneurs, and left for Dublin in March 19, 2000. We quickly reviewed the trash project and decided we could probably contribute to the success of the project. Soon an excellent rapport was developing between the parties, so the president introduced us to the ex-prime minister of Ireland, Mr. Albert Reynolds, TD.

You may be wondering why I bothered to introduce the Irish trash project to a book about airplane crashes. There was a very significant reason. On April 13, 2000 we had been invited back to Ireland to continue with the business discussions. We set up shop in the Clontarf Castle Hotel and began the normal business "due diligence", which was going very slowly as the management seemed more interested in working with some Russians than solving the Irish trash problem.

On April 17, 2000, I got up unusually early and was wondering if we were ever going to get down to business with the Irish folks. I gathered the morning papers, and sat down with a coffee in the breakfast room. I had only been there for a coffee or two when the Irish folks came in, with four or five Russian diplomats and businessmen, together with Albert Reynolds. Suddenly, I was invited to the breakfast and introduced as an American entrepreneur working on the trash project in Ireland. The conversation was half Russian and half Irish and was clearly set up to introduce Albert Reynolds to this potentially huge Russian business for Ireland.

As the meeting broke up, Reynolds pulled me to one side an asked me to stay for a quick chat — a chat that lasted one and a half hours. I was fascinated. Why did this opportunity to learn more about Pan Am 103 come to me when I might easily have gone home, upset by the delays in the Irish trash project? Reynolds was very relaxed as he told me how he had met Gadhafi in 1980 and had remained in contact with him over the years, including the period after the downing of Pan Am 103. He went on to say that he had been negotiating with Gadhafi and President Bill Clinton to have the bombing suspects released to Holland to face the criminal charges.

Reynolds continued to say that he advised Gadhafi to let the two suspects stand trial, as he had been assured that each would receive a fair trial. Gadhafi was very concerned that there would be a railroad job that would end up with the crippling UN and American sanctions remaining in place.

I said very little, for my own part, and I told Albert nothing about my investigation. Then the real reason for the visit became evident. For Albert's help with the Libyan negotiations, he had been given the opportunity to develop an oil concession in Libya and was looking for financing. He must have found out that I was a petroleum engineer with experience in oil and gas development in Libya and with connections to oil and gas exploitation funds. Unfortunately, I was unable to help, as America had maintained the sanctions against Libya even though the United Nations had lifted them once the two suspects had arrived in Holland. I made it clear that we might well be interested once the sanctions had been lifted. I knew that Libyan oil was of excellent quality and keenly sought by many companies and countries. Meanwhile, the press had reported that President Mandela had negotiated the release of the two suspects. Maybe Mandela was politically more acceptable as a leader in Africa, and it has been well publicized that Gadhafi had supposedly funded the IRA. I was beginning to wonder just how many groups and individuals had been working to resolve the Pan Am 103 impasse.

The two defendants, Abdelbaset Ali Mohamed Al Megrahi and Al Amin Khalifa Fhimah, had been in jail since April 6, 1999. It is not clear why the two defendants had been detained for over a year.

These are Trying Times

The trial was scheduled to begin on February 2, 2000 but was soon delayed three months by the defense team as it successfully argued that the prosecution had had a decade to prepare, whilst the defense had had only a few months.

Meanwhile, some of the international TV networks were trying to obtain permission to televise the whole court proceeding and were getting nowhere. The BBC filed a lawsuit but still failed. There was a concern that the whole trial would become a circus and politicized. Dr. Swire was very much against TV coverage, but I worried that the trial was getting too little coverage in all media outlets. Hundreds of Americans had died in the Pan Am 103 crash, over $100 million was to be spent by the British and American taxpayers, yet neither was getting the benefit of following the rapidly unfolding events.

The Lockerbie Trial

Without going over all the details of the trial, we should review a few of its aspects. First, in any lawsuit in the Western world, one is usually presumed innocent until proven guilty. If the sanctions against Libya were based on the Pan Am 103 bombing, why were they imposed before at least one of the defendants had been proven guilty?

Why is the Lockerbie trial so mired in state secrets? The TWA 800 case appeared to be handled openly, even though terrorism was never ruled out and the case is still open. Why has the United States Congress asked the Secretary of State for all of the documentation on the negotiations between the United Nations and Libya — and been refused?

Under Scottish law it is incumbent on the prosecution to prove the defendants guilty beyond a reasonable doubt, whereas the defense has no obligation whatsoever to prove innocence. This is a very important legal understanding and its implications would be used to the extreme.

The news media have been little to no help in keeping up with events; I recommend three main web pages that appear to be balanced and well organized: www.geocities.com/CapitolHill/5260 and all its links; www.thelockerbietrial.com, by Professor Robert Black; http://vialls.homestead.com, an excellent independent analysis by Joe Vialls in Australia.

Let me now simply summarize some of the most interesting aspects of the trial as it unfolds, together with other aspects of this investigation its ramifications and developments. The most logical approach will be to discuss the trial chronologically.

Week One (Ending May 7, 2000). The first week was focused mainly on describing the events of the crash in minute detail. The biggest news was a sixteen-page report submitted to the Scottish Crown Office/ Lord Advocate placed on the Internet by MEBO's Ed Bollier. Bollier had hired a world renowned German explosives team under Dr. Hitmar Schubert at the Fraunhofer-Institute, Munich, Germany to investigate whether or not the explosion could have taken place in a Bombeat Cassette player, inside a suitcase, inside a luggage container. It transpired during the trial that one of the prosecution experts admitted making calculation errors in determining the proximity of the container to the side of the plane necessary to cause the damage seen to the wall of the aircraft. Dr. Schubert's study clearly demonstrated that the explosive material actually would have had to be attached to the fuselage. This report became a major concern to the prosecution as their entire case was based on the original scenario.

The court decided to ignore the new evidence and Bollier went ballistic. There was more to come in the weeks ahead.

Week Two (Ending May 14, 2000). This week was again mainly concerned with presenting evidence and it became very clear that much of the evidence had been poorly tagged, poorly documented and even changed. The week was not going well for the prosecution, who therefore requested an adjournment, which was reluctantly approved by the presiding judge. The trial was delayed until May 23, 2000.

Week Three (Ending May 21, 2000). The court remained in adjournment.

Week Four (Ending May 28, 2000). On May 25, Senior Air Accident Board Inspector (AAIB) Christopher Protheroe admitted in court that a complicated formula used to calculate the trajectory of the blast had been incorrectly applied in the case.

This was another blow to the prosecution's case. The revised calculation showed that the explosion had to have been 12 inches from the side of the fuselage, putting it outside the container, rather than 25 inches from the side, which could have been inside the container. The bomb could not have been inside the container; thus, the case against the Libyans was falling starting to unravel.

On May 21, 2000, the *Sunday Herald* exposed yet another bombshell. It brought to the attention of the court the story of a Professor Andrew Fulton, a former MI6 Head of Station in Washington, DC who had been dismissed from the Lockerbie Trial Unit — another story that will not fit within the covers of this book.

On May 30, 2000, there was another postponement while the alleged container, AVE 4041 PA, was dismantled because it would not pass through the door into the courtroom. Bollier told me that he was almost certain that this container was not the one from Pan Am 103. He pointed out that such a container would have been crushed, having hit the ground from an altitude of 30,000 feet. Considering the damage to the entire aircraft, it would be amazing if the container had remained in such a relatively undamaged state. Bollier suggested that this container had been used to conduct a test on the ground.

Week Five (Ending June 4, 2000). The container had been placed in the courtroom and another AAIB inspector described how he had found a small fragment of the circuit board "wedged into a warped metal tag, which is normally on all containers and identifies the manufacturer of the unit". Interestingly enough, as pointed out in the Joe Vialls web page, the ID tag is on the outside of the container and one has to question how the fragment managed to become lodged there, instead of being blasted away. Clearly, this is one more piece of evidence to demonstrate that the explosion was outside the container.

The trial continued questioning one expert after another. The more evidence is given, the more questionable the bomb theory becomes, overall. Maybe the Pan Am 103 was the caused by the same problem with the TWA 800 and Air India 182.

Then came another surprise! On June 3, 2000 the CBS 60 *Minutes* program suggested that the CIA was interviewing a Mr. Ahmad Behbahani (who was in protective custody in Turkey). Behbahani apparently claimed that Iran was behind the bombing of Pan Am 103 and that he had ordered the attack — as revenge for the downing of the Iranian Airbus by the USS Vincennes.

The whole program appeared to have been aired without a complete analysis of the situation. The Behbahani story was dropped from consideration by the court and probably rightly so. The man was only in his early thirties at the time of the interview and hence would have been far too young to have such a commanding position in the Iranian government. One has to wonder why all of these stories were coming to the forefront now that the trial was underway. Some people, like Bollier, have had their lives threatened.

Week Six (Ending June 11, 2000). This should have been another bad week for the prosecution. One of the DERA scientists, John Douse, admitted that no tests for explosives had been conducted on any parts of the MEBO timer, the Toshiba cassette player or even the suitcase.

Curiously enough, the police had discovered very little of the timer or the radio, and yet the whole trial was based on this evidence. Mr. Douse claimed that the explosive tests had not been conducted — to save money. Save money? This entire investigation had cost tens of millions of dollars. The whole trial was looking very shaky and Bollier was instructed to be prepared to bring his research to the witness stand very shortly.

Week Seven (Ending June 18, 2000). The prosecution placed Mr. Alan Feraday on the stand. Mr. Feraday, another DERA scientist, brought to the court a so-called "reconstructed" Toshiba RT SF16 Bombeat radio cassette made of black plastic. His original report had indicated that the plastic explosive material had been in a white Toshiba 8016 or 8026 recorder.

His work was deemed to be worthless.

Later in the week the prosecution brought on Mr. Erwin Meister, Cofounder of MEBO, and grilled him about the sale of the MST13 timers to both Libya and the East German Stasi. Both Meister and Bollier were heading for a bad week.

Week Eight (Ending June 25). Much to the surprise of Mssrs Meister and Bollier, they came under attack by both the prosecution and the defense lawyers. The defense attorneys had decided to take a huge risk by attacking the credibility of the MEBO management. The opportunities this opened to the prosecution were endless. Making the MEBO analysis look fabricated would lend credibility to the poor work by the forensic scientist and the other prosecution investigators. Clearly the defense strategy was to confuse the issues, to try to place the blame on the Palestinians and otherwise create a situation wherein the prosecution would be unable to prove the indicted men guilty beyond a reasonable doubt.

Bollier should have made a formal presentation, with all of his detailed studies, photographs, videos, diagrams, schematics and the other solid evidence he had collected over the previous eight or nine years. An independent qualified technical person, whose native language was English, might even have presented this evidence to ensure a clear understanding. This was not the case and I believe the judges were left with a limited understanding of the truth. The judges can only rule on the evidence they hear and see in court and this might be a tragedy for the accused and the victim's families, who, one must hope, sought nothing but the truth.

Both the prosecution and defense lawyers concentrated on Bollier's previous sales to the East German Stasi and other electronic equipment to Libya. They were attempting to prove to the court that these two individuals were of low (even treasonable) character for having dealt with the East Germans during the Cold War period.

What the court did not hear was that Bollier had registered all of his foreign sales to the Swiss authorities, including the Swiss police, without being challenged at the time. That is what he told me. He even invited me to see all of this evidence once the trial was complete. He showed me a lot of evidence that had not been shown to the court, at this point in time, and it is doubtful that Bollier will be called back to testify.

Week Nine (Ending July 2, 2000). Bollier blew up. He demanded that the defense lawyers be replaced. Bollier had been savaged by the court and he was coming out fighting. First, he issued letter to the Crown Office dated June 28, 2000 entitled, "MEBO-APPEAL. Lockerbie-Appraisal and Proposal." Bollier insisted that the folks who were conducting the forensic work had clearly demonstrated deceit, conspiracy, and intent to cover-up. He insisted that Professor Dr. Hitman Schubert had proven completely that the explosion had taken place outside the AVE 4041 PA container and therefore the Libyans were not involved.

Bollier went on to demand that new tests be conducted to prove whether the bomb must have been inside or outside the container. He stated that the cost of such work would be minimal compared to all of the expenses encountered by the prosecution in gathering the most minute information while ignoring the factual, repeated data. Are lawyers technically competent to consider such Bollier data?

On June 30, 2000, another letter was sent to the Crown Office, Lord Advocate, Scottish Court Camp Zeist entitled, "Lockerbie-Scandal in Camp Zeist!" Next, Bollier had his lawyers announced that they intended to file a criminal suit alleging that the "powers that be" had fabricated evidence, altered and tampered with evidence, and were pursing a course that was detrimental to both the Libyans and himself. In this letter to the court Bollier reviewed all of the evidence presented to him during his time on the stand. While he was in the witness box he asked to see the fragments once more and, to his horror, they had been changed again.

This time he compared the fragments he was seeing in court with the ones he had seen in Scotland on September 15, 1999. It was very clear to him that the fragments had been burned, since the color of the board was neither green nor brown. Further details on this point are to be found on the Internet at Geocities.com/CapitolHill/5260 edbol30600.html.

During the questioning of Mr. Fereday, it transpired that Fereday had taken the so-called Lockerbie fragment to Thurman, whereas Thurman had told *60 Minutes* that he had only seen photographs. Thurman indicated at the same time that he thought the Scottish police had cut the fragment, which would have explained why the fragment was different from the commercially-made circuit boards sent to Libya.

This evidence really did require a very careful review from the very beginning. Bollier made it very clear that his lawyers were about to file a criminal lawsuit to find the truth — and the truth is what we all must seek, if we are to remain a free society.

Then came the testimony of Mr. Ueli Lumpert, a former employee of MEBO. I do not see the value of expressing Mr. Lumpert's testimony, as he appeared to have memory problems. However, his testimony has been reviewed by Professor Black and can be reviewed on the Internet at: www.thelockerbietrial.com/trial_news_june28.html.

The court then recessed until July 11, 2000.

Mr. Bill Curtis presented an exposé on the well-known TV program *Investigative Reports*, called "Friend or Foe"; it offered a damning review of what actually happened aboard the warship Vincennes when it destroyed the Iranian airbus over the Straits of Hormuz. Curtis's main points were that:

- The Vincennes fired on Iran Air Flight 655 on July 1988, resulting in the death of 290 innocent passengers, while it was cruising in Iranian territorial waters — a clear violation of international law and its "rules of engagement." A billion-dollar, high-tech Aegis cruiser, the Vincennes was known by the other warships in the area as "Robocruiser", for its aggressive behavior in such situations

- The Vincennes was a very advanced warship capable engaging up to 200 incoming missiles at any one time and was supposedly engaged in a spat with some Iranian gunboats, called "Boghammers", which were barely twenty feet long. This would be like shooting squirrels with a laser-guided missile. The fight appeared to have been provoked when Captain Rogers sent a helicopter to check out the Iranian gunboats. The helicopter radioed that it may have come under fire as it observed several white flashes from the area near the gunboats. These white flashes might easily have been reflections in the sea, as the gunboats churned up the water.

At about the same time, Iran Air's Airbus, Flight 655, was taking off from Bandar Abbas Airport on a routine flight to Dubai. Unfortunately, the flight path was directly over the Vincennes, which almost immediately assumed it was under attack. The fact that the Airbus was slow and was climbing, unlike a fighter jet, was ignored; the plane was shot down. The other U.S. warships in the area were well aware of the real situation but were unable to stop Captain Rogers from taking his fateful action.

A big cover-up ensued. About eleven hours later, Admiral William Crowe, chairman of the Joint Chiefs of Staff, announced the situation to the press (based on reports by Captain Rogers). The Pentagon knew what had happened but deliberately misled both Vice President Bush and indeed the entire U.S. Congress. A more in-depth evaluation was initiated, but the assigned investigator said, according to the Bill Curtis investigation, "It was too embarrassing for the U.S. Congress, and further work was dropped."

Then came the real affront to the Iranian victims and their families. The military awarded Captain Rogers and his crew medals for "Heroic Achievement". The Iranians made it very clear they were going to extract an "Eye for an Eye." Or a plane for a plane?

On July 6, 2000, a story came out about the involvement of Oliver "Buck" Revell, who had filed a $10 million lawsuit against Hart G. W. Lidow, a medical researcher at Harvard School of Medicine. Mr. Lidow was Flora Swire's suitor (Dr. Jim Swire's daughter). Lidow maintains that Oliver Revell knew of the bombing and had his own son change flights. Flora was given his seat assignment and, as Jane Swire told me during one of my visits to their home, "Flora boarded Pan Am 103 like a sheep going to the slaughter".

It is very clear that the U.S. and British authorities are determined to kill this story by whatever means they can. A few significant lawsuits will be files in the USA and UK after the trial, which we must hope will bring to light the real truth.

Week Ten (Ending July 16, 2000). Each side was commenting on the lack of security at both the Malta airport and the Frankfurt airport. The main news was based on the fact that many Maltese witnesses refused to come to the trial. The significance of their potential evidence was debated endlessly and probably added nothing to the understanding of what really happened.

One very significant fact was the admission by the Scottish police that they had illegally wire-tapped many Maltese telephones. This upset the Maltese government, as one might well expect. Privacy is becoming a thing of the past.

Week Eleven (Ending July 23, 2000). The full details of the Bollier criminal complaint was filed by his attorney, Dr. Dieter W. Neupert, and appeared on the Internet. The case was called "Criminal Complaint Regarding the Falsification of Evidence in the Lockerbie Case." This was a powerful accusation, but one notes that it was only a complaint, and not a lawsuit. This complaint could simply be ignored.

The court continued to hear what would have to be irrelevant testimony, about other passengers on the flight from Malta to Frankfurt, who experienced a normal voyage. On his webpage, Joe Vialls make an interesting observation. He suggests that the prosecutors knew that certain Maltese would not show up for the trial, and lots of other circumstantial information is nothing more than attempt by the prosecution to infer guilt because these people are "hiding something". Vialls recommends *Manufacturing Consent,* by Noam Chomsky.

Week Twelve (Ending July 30). On July 23, 2000, Neil Mackay had an article in the *Sunday Herald* entitled "Dangerous Airline List Kept Secret." There were calls for the resignation of Robin Cook, the British Foreign Secretary, over the fact that Tony Blair's government cares more about international business than the welfare of its British travelers.

Then Mr. John Parks, an explosives engineer with some 36 years' experience, submitted a report to the court wherein he claims that the bomb could not have been inside a radio, inside a suitcase, inside a container, and cause the injuries that were sustained by certain passengers. The evidence is compelling and leads once again to the question — was there a bomb at all? The details of Mr. Parks's analysis are posted on the Internet for all to see.

On July 25, 2000, a French supersonic Concorde crashed on takeoff from Charles DeGaulle Airport, near Paris. Remarkably, there were many amateur photographers in the area and the spectacular crash was videoed during its fateful dive into a small hotel. There was clear photographic evidence of a huge fire emanating from an area between the two port-side engines and the fuselage. That did not deter dozens of reporters from mouthing an official version that debris from the runway had been ingested into the engines, causing engine failure. The fact that burning fuel from one of the ruptured tanks was pouring out profusely, thus starving the engines of fuel at a critical moment, was ignored. However, the public was not fooled and all of the Concordes were grounded until the known problems are resolved.

On July 28, 2000, the court recessed for vacation, while the two Libyans, innocent until proven guilty, still languished in jail. The prosecution and defense attorneys meanwhile informed the court that a deal had been made and the trial would not take a year to complete. Indeed, they hoped they might even finish within a few months.

On July 31, Cliff Kincaid (with the Washington, DC-based watchdog organization "Accuracy in the Media") introduced an article entitled "TWA 800 Eyewitnesses Demand a Hearing." Mr.

Kincaid is no friend of Libya, what with his unsubstantiated claims that Libya was responsible for Pan Am 103 long before the trial even started. He is correct when he states that many inquisitive people do not believe the official version of events surrounding TWA 800. I have tried, with no success, to get a copy of the video from AIM in which an amateur attempts to ignite a container of jet fuel.

Weeks 13, 14, & 15 (Ending August 19, 2000). The trial was in recess. On August 2, 2000, *USA Today* reported that Boeing had been fined $1.2 million for quality-control problems. Boeing was accused of many infractions, such as applying heat treatment to parts at the wrong stage of production, failure to keep machine tools precisely calibrated, and not notifying the FAA about safety issues reported by airlines for over a year.

On August 16, 2000, the *International Herald Tribune* reported that the "TWA 800 Report Won't Solve the Mystery." In this article is a discussion of the tests conducted on a one-quarter sized model fuel tank, which was supposed to simulate the explosion that was considered to have brought down TWA 800. Bingo — they finally admitted that instead of using the same jet fuel as in the TWA 800 flight for their experiment they had used propane, which is infinitely more explosive than jet fuel. This test was shown *ad nauseum* on CNN, with a small test tank blowing up with a huge fireball blasting from the right hand side of the tank. It was spectacular, but irrelevant. The fire was far too bright, with too little black smoke — which means it was not the heavier fuel — the jet fuel.

Then, *on August 17,* the same paper reported that airline fuel tanks had been found safe. In fact, these tanks have been safe for decades and all pilots, engineers, FAA people, and airlines folks knew it. If there had been any real question of airplane fuel tanks exploding, the planes would have been grounded.

There is an excellent webpage available for the curious, which has been prepared by William S. Donaldson retired USN at www.twa800.com. Here you can download the video of the backyard fuel explosion test conducted by the commander. This website is very thorough and controversial. The Commander is completely convinced that the NTSB and the many other agencies of the Clinton administration are covering up the truth. He and many others believe that a missile shot down TWA 800. Whereas, I might agree with the ultimate conclusion there is precious little concrete evidence that a missile brought down this plane. I would rather believe that it had been hit with a HERF gun. If you wish to know what a HERF gun is I suggest you read Mr. Winn Schwartau's excellent book "Information Warfare." You might also be interested to know that a detailed documentary conducted by the Swiss German TV, DTV, indicated that a HERF gun probably brought down the Swissair 111, which crashed near Peggy's Cove, Nova Scotia, Canada. This documentary has never been seen in English in the USA. There is a reason! An Algerian businessman close to Clinton was on the plane — but that's another story for another book. These guns have been used against drug smugglers! I will not elaborate since I believe that these Boeing 747's crashes were caused by metal fatigue of poor quality materials — as you have no doubt already gathered.

Week Sixteen (Ending September 2, 2000). On August 22, the Lockerbie trial started what was supposed to be the big climax for the prosecution. It turn into another major fiasco. The prosecution wanted to bring on a Libyan defector, Abdul Majid Razkaz Salam Giaka, who had been a CIA informant at the time that the two Libyans on trial were accused of bombing Pan Am 103. He had been on the CIA payroll and kept in the USA under the witness protection program.

A series of CIA transcripts, telegrams and other documents were presented to the court, but they had been heavily redacted. The defense protested — the prosecution had seen the original documents and the defense demanded to see them too. Some additional documents were later provided, which brought to the attention of the defense team that other important documents must also available. To make a long story short, the court was once again placed in recess until September 21, 2000. This was a serious blow to the prosecution, which had not presented any conclusive evidence thus far.

On the same day, the NTSB put on yet another public hearing wherein it declared that the TWA 800 crash inquiry was now closed. It blamed the entire incident on an explosion in the Boeing 747's central fuel tank. Few people believed a word. Interestingly enough, Mr. Reed Irvine, president of Accuracy in the Media, was forcibly rejected from the hearings as he tried to present the hearing with his version of the missile theory.

Meanwhile, Mr. Bollier was not silent and one can understand why. He sent out a message to all explosives experts to help him get his data presented before the court. Since neither the prosecution nor the defense wanted anything to do with Bollier, the chances of him being heard were very slim.

During one of my interviews with Bollier, he told me that the Iran Air terminal was next to the Pan Am terminal at Heathrow — Pan Am 103's last stop before Lockerbie. This was confirmed this week at the hearings, which means that someone could have placed the bomb on the plane at Heathrow. That is much more logical and would confirm Bollier's expert's conclusion that the bomb was placed against the side of the plane and not in a radio, in a suitcase in a container in the aircraft hold — if there had been a bomb at all.

On August 31, 2000, Mike Wallace on the History Channel discussed both the Pan Am 103 and the Korean Air 007 crashes. He offered nothing but propaganda, which Mike Wallace has been accused of many times.

Week Seventeen (Ending September 9, 2000). The trial was in recess until September 21, 2000. However, this did not deter Ed Bollier from posting a scathing Internet attack on both the prosecution and defense with a twenty-point analysis of events. Regardless of the outcome of this trial, one can be assured that Bollier will not be silent!

October and November 2000. There were some very significant developments in both the Lockerbie trial and the TWA 800 investigation during this period. The Lockerbie trial was repeatedly delayed by new evidence and other legal maneuvers.

On September 26, 2000, a former CIA mole, a 40-year-old Libyan identified as Abdul Majid Abdul Razkaz Abdul-Salam Giaka took the stand at the custom-made Scottish court in Zeist. Giaka was the prosecution's key witness, as he had claimed to have seen the two accused Libyans place the bomb on the Malta flight, which linked up with Pan Am 103. Giaka had been a CIA informant drawing $1,000 a month for his efforts. The CIA soon learned that he was not as close to the Libyan regime as claimed and was cut off. Later he was spirited out of Malta by the CIA, against Maltese law, and placed under the U.S. witness protection program where he remained for over ten years. Gala was a major disappointment for the prosecution; the Crown's case was coming unglued.

Through these two months, Ed Bollier sent letters to the Scottish court, demanding to be heard. He appealed to the world for lawyers to enter the fray in the name of justice. The Internet was loaded with detailed analyses of events — but they appeared to be falling on deaf ears. The press and the international TV stations were ignoring the "Trial of the Century".

On October 7, 2000, the author visited a director and producer of documentaries in Washington DC to explore the possibility of making a series of TV programs on this subject. During the meetings, mention was made of a group of people who were demanding that the U.S. Congress look into the downing of TWA 800 by a missile. (Several hundred people claim to have witnessed the missile heading towards the doomed plane.) While the author believes these folks are barking up the wrong tree, this intensity of interest clearly demonstrates that the informed public does not buy the official version of what happened to TWA 800. The website at twa800.com provides excellent details on this troubling fiasco.

On October 9, 2000, this ever-convoluted trial suffered yet another amazing — new evidence of considerable sensitivity — and the trial went on hold for yet another week. Syria was back in the limelight.

On October 12, 2000, the USS Cole, an Aegis Class guided missile destroyer was in the port of Aden on a refueling mission, en route from the Red Sea to Bahrain, when it was bombed.

On October 16, 2000, Seth Cropsey with the *Washington Times* presented an article entitled "War, Not Terrorism". Around the world, nations and indeed individuals are in a state of guerrilla warfare against the USA and Europe. Guerrilla war — because frustrated counties cannot air their grievances against the most powerful nations in the world on the conventional battlefield.

On October 27 2000, the Royal Canadian Mounted Police arrested two men in connection with the alleged bombing of Air India 182.

On October 31, 2000, *USA Today* had an article entitled, "FAA Probe Reveals Flaws at Boeing — High Visibility Breakdowns". Could this have been the case while the first 600 Boeing 747's were being assembled?

November 7, 2000. The Pan Am trial was bogged down behind closed doors. The court was tied up with conflicting versions of the original story that the Palestinians were behind the bombing of Pan Am 103. Ahmed Jabril, Mohammed Abu Talb and now a Balkans connection were further complicating the story. On November 10, 2000, it was Abu Talb's turn on the courtroom stand; the same Abu Talb who was jailed for life in Sweden having been convicted of the terrorist act of bombing a synagogue in Denmark in 1989. In a Scottish court the defense has no obligation to prove innocence — it only has to provide the three judges with sufficient proof that these two Libyans cannot be convicted "Beyond a Reasonable Doubt". The prosecution told the court that it expected to rest its case in early December. The case dragged on but the story will not go away that easily.

By November 20, 2000, the prosecution had completed its case and the defense for Fhimah immediately insisted that insufficient evidence had been presented and demanded that Fhimah be released. The defense for Magrahi decided it would continue with the defense arguments but requested more time to analyze recently acquired information — the request was approved and the trial, once more, was delayed until December 5, 2000 but not before the Zeist court refused to release Fhimah.

The Pan Am 103 trial will be concluded by the time you read this book. I expect that the two Libyans will be set free, which will be a fiasco. The victims' families will demand further inquiries. Ed Bollier will no doubt file a major lawsuit against those who have caused him so much grief and the story will continue for years to come.

The TWA 800 crash will not easily leave the radar screen as the Accuracy-in-Media group in Washington DC will continue to demand that the US Congress get involved to further investigate their hypothesis that TWA 800 was brought down by a missile.

The Air India 182 trial will soon begin, which will bring additional attention to these series of events.

People with more information would be forthcoming and explain the circumstances under which these planes have crashed.

But for now, I believe Franck Curtin's "Boeing Report" holds the most plausible, and in some ways the most disturbing, explanation for so many of these tragedies.

SELECTIVE BIBLIOGRAPHY

The following books and references will add to your understanding of the basis and conclusions exposed in this book:

World Directory of Airlines Crashes. Terry Deaham, ISBN 1 85260 554 5

Air Disasters' Dialogue from the Black Box. Stanley Stewart, ISBN 1 56619 671x

Air Disasters. Stanley Stewart, ISBN 1 56619 671x Copy Number 2

Legend and Legacy. Boeing and its People. Robert J. Sterling, ISBN 0 312 05890 x

Collision Course. Ralph Nader, ISBN 0 8306 4271 4

The Passionate Attachment. George and Douglas Ball, ISBN 0 393 02933 6

Libya — Dessert In Conflict. Ted Gottreid, ISBN 1 56294 351 0

Incident over Sakhalin. Michel Brun, ISBN 1 56858 054 1

Flying Blind, Flying Safe. Mary Schiaro, ISBN 0 380 97532 7

The Fall of Pan Am 103 — Inside the Lockerbie Investigation Steven Emerson & Brian Duffy, ISBN 0 399 13521 9

The Trail of the Octopus. From Beirut to Lockerbie. Inside the DIA. Donald Goddard & Lester K. Coleman, ISBN 0 7475 2562X

Full Disclosure. Andrew Neil, ISBN 0 333 64682 7

Raid on Gadhafi. Col. Robert E. Venkus, ISBN 0 312 07073X

Sky Fever. Sir Geoffrey de Havilland No, ISBN

The Downing of Flight TWA 800. James Sanders, ISBN 08217 5829 2

Information Warfare. Chaos on the Information Highway. Winn Schwartau, ISBN 1560251328

The Crimes of a President. New Revelations, Conspiracy & Cover up by the Bush and Reagan Administration. Joel Bainerman, ISBN 1561711888

Pan Am 103. The Bombing, the Betrayals and a Bereaved Family. Search for Justice. Susan Cohen, June 2000,, ISBN 0 451 20165 5

Pan Am 103. The Lockerbie Cover Up 4221. William C. Chasey, ISBN 0 9640104 1 0

Sell Out. The Inside Story of President Clinton's Impeachment — The Most Corrupt President in US History. David Shippers., ISBN 0895262436

A Vast Conspiracy. Jeffrey Tooban & Bill Bittman, ISBN 0743204131

Year of the Rat. How Bill Clinton Compromised US Security for Chinese Money. Edward Timperlake & William C Triplett

Betrayal. Bill Gertz, ISBN 0 89526 317 3

Uncovering Clinton. Michael Isikoff, ISBN 0 609 60393 0

Dollars for Terror. Richard Labeviere. English Version Contact Algora Publishing. New York.

Compromised. Clinton, Bush and the CIA. Terry Reed & John Cummings ISBN 1 883955 02 5

The Media Disasters Pan Am 103. Joan Deppa, Maria Russell, ISBN 0 8147 1856 6

Recommended Internet Websites:

www.planetruth.com. The webpage for this book
www.geocities.com/CapitolHill/5260/
http://headlines.yahoo.com/pan_am_103_Lockerbie_Trial/
http://headlines.yahoo.com/Full_Coverage/US/TWA_Flight_800/
www.TWA800.com

Also from Algora Publishing:

CLAUDIU A. SECARA
THE NEW COMMONWEALTH
From Bureaucratic Corporatism to Socialist Capitalism

The notion of an elite-driven worldwide perestroika has gained some credibility lately. The book examines in a historical perspective the most intriguing dialectic in the Soviet Union's "collapse" — from socialism to capitalism and back to socialist capitalism — and speculates on the global implications.

IGNACIO RAMONET
THE GEOPOLITICS OF CHAOS

The author, Director of Le Monde Diplomatique, presents an original, discriminating and lucid political matrix for understanding what he calls the "current disorder of the world" in terms of Internationalization, Cyberculture and Political Chaos.

TZVETAN TODOROV
A PASSION FOR DEMOCRACY –
Benjamin Constant

The French Revolution rang the death knell not only for a form of society, but also for a way of feeling and of living; and it is still not clear what we gained from the changes.

MICHEL PINÇON & MONIQUE PINÇON-CHARLOT
GRAND FORTUNES –
Dynasties of Wealth in France

Going back for generations, the fortunes of great families consist of far more than money— they are also symbols of culture and social interaction. In a nation known for democracy and meritocracy, piercing the secrets of the grand fortunes verges on a crime of lèse-majesté . . . Grand Fortunes succeeds at that.

CLAUDIU A. SECARA
TIME & EGO –
Judeo-Christian Egotheism and the Anglo-Saxon Industrial Revolution

The first question of abstract reflection that arouses controversy is the problem of Becoming. Being persists, beings constantly change; they are born and they pass away. How can Being change and yet be eternal? The quest for the logical and experimental answer has just taken off.

JEAN-MARIE ABGRALL
SOUL SNATCHERS: THE MECHANICS OF CULTS

Jean-Marie Abgrall, psychiatrist, criminologist, expert witness to the French Court of Appeals, and member of the Inter-Ministry Committee on Cults, is one of the experts most frequently consulted by the European judicial and legislative processes. The fruit of fifteen years of research, his book delivers the first methodical analysis of the sectarian phenomenon, decoding the mental manipulation on behalf of mystified observers as well as victims.

Jean-Claude Guillebaud
THE TYRANNY OF PLEASURE

Guillebaud, a Sixties' radical, re-thinks liberation, taking a hard look at the question of sexual morals -- that is, the place of the forbidden -- in a modern society. For almost a whole generation, we have lived in the illusion that this question had ceased to exist. Today the illusion is faded, but a strange and tumultuous distress replaces it. No longer knowing very clearly where we stand, our societies painfully seek answers between unacceptable alternatives: bold-faced permissiveness or nostalgic moralism.

Sophie Coignard and Marie-Thérèse Guichard
FRENCH CONNECTIONS –
The Secret History of Networks of Influence

They were born in the same region, went to the same schools, fought the same fights and made the same mistakes in youth. They share the same morals, the same fantasies of success and the same taste for money. They act behind the scenes to help each other, boosting careers, monopolizing business and information, making money, conspiring and, why not, becoming Presidents!

Vladimir Plougin
RUSSIAN INTELLIGENCE SERVICES. Vol. I. Early Years

Mysterious episodes from Russia's past – alliances and betrayals, espionage and military feats – are unearthed and examined in this study, which is drawn from ancient chronicles and preserved documents from Russia, Greece, Byzantium and the Vatican Library. Scholarly analysis and narrative flair combine to give both the facts and the flavor of the battle scenes and the espionage milieu, including the establishment of secret services in Kievan Rus, the heroes and the techniques of intelligence and counter-intelligence in the 10th-12th centuries, and the times of Vladimir.

Jean-Jacques Rosa
EURO ERROR

The European Superstate makes Jean-Jacques Rosa mad, for two reasons. First, actions taken to relieve unemployment have created inflation, but have not reduced unemployment. His second argument is even more intriguing: the 21st century will see the fragmentation of the U. S., not the unification of Europe.

André Gauron
EUROPEAN MISUNDERSTANDING

Few of the books decrying the European Monetary Union raise the level of the discussion to a higher plane. European Misunderstanding is one of these. Gauron gets it right, observing that the real problem facing Europe is its political future, not its economic future.

DOMINIQUE FERNANDEZ

PHOTOGRAPHER: FERRANTE FERRANTI

ROMANIAN RHAPSODY — *An Overlooked Corner of Europe*

"Romania doesn't get very good press." And so, renowned French travel writer Dominique Fernandez and top photographer Ferrante Ferranti head out to form their own images. In four long journeys over a 6-year span, they uncover a tantalizing blend of German efficiency and Latin nonchalance, French literature and Gypsy music, Western rationalism and Oriental mysteries. Fernandez reveals the rich Romanian essence. Attentive and precise, he digs beneath the somber heritage of communism to reach the deep roots of a European country that is so little-known.

PHILIPPE TRÉTIACK

ARE YOU AGITÉ? *Treatise on Everyday Agitation*

"A book filled with the exuberance of a new millennium, full of humor and relevance. Philippe Trétiack, a leading reporter for Elle, goes around the world and back, taking an interest in the futile as well as the essential. His flair for words, his undeniable culture, help us to catch on the fly what we really are: characters subject to the ballistic impulse of desires, fads and a click of the remote. His book invites us to take a healthy break from the breathless agitation in general." — Aujourd'hui le Parisien

"The 'Agité,' that human species that lives in international airports, jumps into taxis while dialing the cell phone, eats while clearing the table, reads the paper while watching TV and works during vacation – has just been given a new title." — Le Monde des Livres

PAUL LOMBARD

VICE & VIRTUE — *Men of History, Great Crooks for the Greater Good*

Personal passion has often guided powerful people more than the public interest. With what result? From the courtiers of Versailles to the back halls of Mitterand's government, from Danton — revealed to have been a paid agent for England — to the shady bankers of Mitterand's era, from the buddies of Mazarin to the builders of the Panama Canal, Paul Lombard unearths the secrets of the corridors of power. He reveals the vanity and the corruption, but also the grandeur and panache that characterize the great. This cavalcade over many centuries can be read as a subversive tract on how to lead.

RICHARD LABÉVIÈRE

DOLLARS FOR TERROR — *The U.S. and Islam*

"In this riveting, often shocking analysis, the U.S. is an accessory in the rise of Islam, because it manipulates and aids radical Moslem groups in its shortsighted pursuit of its economic interests, especially the energy resources of the Middle East and the oil- and mineral-rich former Soviet republics of Central Asia. Labévière shows how radical Islamic fundamentalism spreads its influence on two levels, above board, through investment firms, banks and shell companies, and clandestinely, though a network of drug dealing, weapons smuggling and money laundering. This important book sounds a wake-up call to U.S. policy-makers." — *Publishers Weekly*

Jeannine Verdès-Leroux
DECONSTRUCTING PIERRE BOURDIEU
Against Sociological Terrorism From the Left

Sociologist Pierre Bourdieu went from widely-criticized to widely-acclaimed, without adjusting his hastily constructed theories. Turning the guns of critical analysis on his own critics, he was happier jousting in the ring of (often quite undemocratic) political debate than reflecting and expanding upon his own propositions. Verdès-Leroux has spent 20 years researching the policy impact of intellectuals who play at the fringes of politics. She suggests that Bourdieu arrogated for himself the role of "total intellectual" and proved that a good offense is the best defense. A pessimistic Leninist bolstered by a ponderous scientific construct, Bourdieu stands out as the ultimate doctrinaire more concerned with self-promotion than with democratic intellectual engagements.

Henri Troyat
TERRIBLE TZARINAS

Who should succeed Peter the Great? Upon the death of this visionary and despotic reformer, the great families plotted to come up with a successor who would surpass everyone else — or at least, offend none. But there were only women — Catherine I, Anna Ivanovna, Anna Leopoldovna, Elizabeth I. These autocrats imposed their violent and dissolute natures upon the empire, along with their loves, their feuds, their cruelties. Born in 1911 in Moscow, Troyat is a member of the Académie française, recipient of the Prix Goncourt.

Jean-Marie Abgrall
HEALING OR STEALING — Medical Charlatans in the New Age

Jean-Marie Abgrall is Europe's foremost expert on cults and forensic medicine. He asks, are fear of illness and death the only reasons why people trust their fates to the wizards of the pseudo-revolutionary and the practitioners of pseudo-magic? We live in a bazaar of the bizarre, where everyday denial of rationality has turned many patients into ecstatic fools. While not all systems of nontraditional medicine are linked to cults, this is one of the surest avenues of recruitment, and the crisis of the modern world may be leading to a new mystique of medicine where patients check their powers of judgment at the door.

Dr. Deborah Schurman-Kauflin
THE NEW PREDATOR: WOMEN WHO KILL — Profiles of Female Serial Killers
This is the first book ever based on face-to-face interviews with women serial killers.

Rémi Kauffer
DISINFORMATION — US Multinationals at War with Europe
"Spreading rumors to damage a competitor, using 'tourists' for industrial espionage. . . Kauffer shows how the economic war is waged." — *Le Monde*
"A specialist in the secret services, he notes, 'In the era of CNN, with our skies full of satellites and the Internet expanding every nano-second, the techniques of mass persuasion that were developed during the Cold War are still very much in use – only their field of application has changed.' His analysis is shocking, and well-documented." — *La Tribune*